BUILDING A DIVERSE WORK FORCE

SCIENTISTS AND ENGINEERS IN THE OFFICE OF NAVAL RESEARCH

Committee to Study Diversity in the Scientific and Engineering Work Force of the Office of Naval Research

Committee on Women and Science in Engineering

Office of Scientific and Engineering Personnel

Naval Studies Board

National Academy Press
Washington, D.C. 1997

NOTICE: The project that is the subject of this report was approved by the Governing Board of the National Research Council, whose members are drawn from the councils of the National Academy of Sciences, the National Academy of Engineering, and the Institute of Medicine. The members of the committee responsible for the report were chosen for their special competences and with regard for appropriate balance.

This report has been reviewed by a group other than the authors according to procedures approved by a Report Review Committee consisting of members of the National Academy of Sciences, the National Academy of Engineering, and the Institute of Medicine.

The National Academy of Sciences is a private, nonprofit, self-perpetuating society of distinguished scholars engaged in scientific and engineering research, dedicated to the furtherance of science and technology and to their use for the general welfare. Upon the authority of the charter granted to it by the Congress in 1863, the Academy has a mandate that requires it to advise the federal government on scientific and technical matters. Dr. Bruce M. Alberts is president of the National Academy of Sciences.

The National Academy of Engineering was established in 1964, under the charter of the National Academy of Sciences, as a parallel organization of outstanding engineers. It is autonomous in its administration and in the selection of its members, sharing with the National Academy of Sciences the responsibility for advising the federal government. The National Academy of Engineering also sponsors engineering programs aimed at meeting national needs, encourages education and research, and recognizes the superior achievements of engineers. Dr. William A. Wulf is president of the National Academy of Engineering.

The Institute of Medicine was established in 1970 by the National Academy of Sciences to secure the services of eminent members of appropriate professions in the examination of policy matters pertaining to the health of the public. The Institute acts under the responsibility given to the National Academy of Sciences by its congressional charter to be an adviser to the federal government and, upon its own initiative, to identify issues of medical care, research, and education. Dr. Kenneth I. Shine is president of the Institute of Medicine.

The National Research Council was organized by the National Academy of Sciences in 1916 to associate the broad community of science and technology with the Academy's purposes of furthering knowledge and advising the federal government. Functioning in accordance with general policies determined by the Academy, the Council has become the principal operating agency of both the National Academy of Sciences and the National Academy of Engineering in providing services to the government, the public, and the scientific and engineering communities. The Council is administered jointly by both Academies and the Institute of Medicine. Dr. Bruce M. Alberts and Dr. William A. Wulf are chairman and vice chairman, respectively, of the National Research Council.

This work was performed under Department of Navy Contract N00014-93-C-0089 issued by the Office of Naval research under contract authority NR 201-124. However, the content does not necessarily reflect the position or the policy of the Department of the Navy or of the government, and no official endorsement should be inferred.

Additional copies of this report are available from:
Office of Scientific and Engineering Personnel
National Research Council
2101 Constitution Avenue, N.W.
Washington, DC 20418

Printed in the United States of America

COMMITTEE TO STUDY DIVERSITY IN THE SCIENTIFIC AND ENGINEERING WORK FORCE OF THE OFFICE OF NAVAL RESEARCH

Harrison Shull, *Chair*
Member, National Academy of Sciences

Linda Cain
Director
Office of University and Science Education
Oak Ridge National Laboratory

George Campbell, Jr.
President and CEO
NACME, Inc.

Katharine B. Gebbie
Director
Physics Laboratory
National Institute of Standards and Technology

Eve L. Menger
Director
Characterization Science & Services
Corning, Inc.

Herbert Rabin
Director, Engineering Research Center
Associate Dean, A. J. Clark School of Engr.
University of Maryland College Park

Frank J. Talamantes
Professor of Biology
Sinsheimer Laboratories
University of California, Santa Cruz

Toni Tomacci
Multicultural Diversity Specialist
Director, Human Resources
Matridigm Corporation

Stephen J. Lukasik, *Liaison*
Committee on Women in Science and Engineering

NATIONAL RESEARCH COUNCIL STAFF

Ronald Taylor
Director
Naval Studies Board

Marilyn Baker
Associate Executive Director
Office of Scientific and Engineering Personnel

COMMITTEE ON WOMEN IN SCIENCE AND ENGINEERING

Howard Georgi, *Co-chair*
Mallinckrodt Professor of Physics
Harvard University

Lilian Shiao-Yen Wu, *Co-chair*
Thomas J. Watson Research Center
IBM Corporation

Harold Amos
Maude & Lilian Presley
Professor Emeritus of Molecular Genetics
Harvard Medical School

Anita Borg
Consultant Engineer
Digital Equipment Corporation

Jane E. Buikstra
Professor of Anthropology
University of New Mexico

Deborah Jackson
Member, Technical Staff
Jet Propulsion Laboratory

Frances E. Lockwood
Vice President
Technical & Product Development
Valvoline

Stephen J. Lukasik
Los Angeles, California

Willie Pearson, Jr.
Professor of Sociology
Wake Forest University

Edward Roy, Jr.
Vice President for Academic Affairs
Trinity University

Susan Solomon
Senior Scientist
NOAA

Lois Steele
Acting Division Director
Human Resources Systems Development
Indian Health Service—Tucson

Julia Weertman
Walter P. Murphy Professor
Robert R. McCormick School of Engineering
and Applied Science
Northwestern University

NATIONAL RESEARCH COUNCIL STAFF

Linda C. Skidmore
Director

OFFICE OF SCIENTIFIC AND ENGINEERING PERSONNEL ADVISORY COMMITTEE

M.R.C. Greenwood, ***Chair***
Chancellor
University of California, Santa Cruz

David Breneman
Dean
Curry School of Education
University of Virginia

Nancy Cantor
Dean, Horace H. Rackham
School of Graduate Studies
Vice Provost, Academic Affairs
University of Michigan

Carlos Gutierrez
Professor of Chemistry
California State University, Los Angeles

Stephen J. Lukasik
Los Angeles, California

Barry Munitz
Chancellor
The California State University

Janet Norwood
Senior Fellow
The Urban Institute

John D. Wiley
Provost
Vice Chancellor for Academic Affairs
University of Wisconsin, Madison

Tadataka Yamada
President
Healthcare Services
SmithKline Beecham Corporation

A. Thomas Young
North Potomac, Maryland

William H. Miller, *ex-officio*
Department of Chemistry
University of California, Berkeley

NATIONAL RESEARCH COUNCIL STAFF

Charlotte Kuh
Executive Director

Marilyn J. Baker
Associate Executive Director

NAVAL STUDIES BOARD

NATIONAL RESEARCH COUNCIL STAFF

PREFACE

In late 1994, the Office of Naval Research (ONR) asked the National Research Council (NRC) to provide advice on how to ensure diversity in its future science and engineering work force in order to meet the needs of anticipated naval science and engineering specialties. The NRC Naval Studies Board and the Office of Scientific and Engineering Personnel, through its Committee on Women in Science and Engineering, convened a one-day planning meeting to discuss an appropriate response to this request. At that meeting, Dr. Fred Saalfeld, Deputy Chief of Naval Research, suggested that the NRC might examine (1) how to increase diversity within ONR headquarters and thus indirectly within the Navy Laboratories, (2) how to recruit women, minorities, and persons with disabilities to ONR science and technology positions, (3) how to create a work climate in ONR that permits individuals to reach their full professional potential and to enter management ranks, and (4) how ONR's educational programs can be used to help increase the diversity of the long-term national pool of scientists and engineers, as well as those at ONR.

These discussions led to the creation of a two-part study. Task 1 was to identify short-term activities that ONR might undertake to address recruitment, retention and attrition, utilization, and career development of the women, minority, and disabled scientists and engineers in its own science and engineering (hereafter referred to as S&E) work force; to analyze the current work environment; and to identify examples of programs elsewhere that might be used or modified for use at ONR.

Task 2 was intended to address the longer-term pipeline issues and to examine the effectiveness of ONR's educational programs (its "corporate programs") in promoting diversity in the future national S&E work force and at ONR. The committee was asked to review the origins, objectives, and effectiveness of these programs and to determine how they might be modified to help increase the supply and availability of underrepresented scientists and engineers.

While this study is in response to a request from the Office of Naval Research, the issue of how to increase the diversity of a scientific or technical work force is a generic one. The committee believes most of the recommendations contained in this report have applicability to other federal agencies as well, and many apply to universities and industrial organizations trying to increase the diversity of their science and engineering work force.

A few comments on the methodology and scope of the study are in order. The committee met four times to analyze information and to deliberate. In addition, a subcommittee met separately to examine ONR's corporate programs. The former Chief of Naval Research, the Deputy

Chief of Naval Research, the Chair of the ONR Diversity Committee, and the Deputy Director of the Corporate Programs Division all provided briefings.

The committee used a combination of quantitative and qualitative methods to gather information. Data on the pools of scientists and engineers in the nation's work force potentially eligible for current or future positions at ONR were generated from the 1993 Survey of Doctorate Recipients, the 1990–95 Surveys of Earned Doctorates, and the 1993 National Survey of College Graduates, sponsored by the National Science Foundation (NSF) and other federal agencies. ONR provided extensive data on the demographic, educational, and employment characteristics of the 150 individuals in its S&E group, as well as samples of recent hiring actions.

A consultant to the committee conducted in-depth personal interviews with 71 of the 150 scientists and engineers, 66 of whom also returned a written questionnaire concerning their attitudes towards a number of issues related to diversity. Summaries of the interview results can be found in Appendix C.

The committee has attempted to gather information on successful diversity initiatives in organizations similar to ONR. Although it could not provide a comprehensive overview of such initiatives within the time and budget available, the committee did identify several examples of programs that can be adapted for use at ONR. In reviewing ONR's corporate programs, the committee examined descriptive material, data on participants, and sample program evaluations. It did not conduct an independent evaluation of each program.

Although early discussions with Navy representatives on the scope of the study indicated a desire to address the Navy's scientific work force, subsequent discussions with ONR leadership established that the committee should focus on the scientific and engineering work force in ONR, which consists of approximately 150 professionals housed primarily in its headquarters in Arlington, Virginia, who manage Navy research and development funds. These program officers and their managers administer $1.4 billion annually and serve as a principal interface between the Navy and the academic community. Changing this work force can have an impact throughout the Navy's science and engineering community, including ONR's 5,400 principal investigators.

There are nearly 1,500 scientists and engineers who work in the Naval Research Laboratory (NRL), but they are primarily practicing scientists and engineers, not managers of research funds. Although their qualifications and backgrounds may be very similar to ONR scientists and engineers, their work activities are not. In consultation with ONR, therefore, the committee decided to exclude NRL personnel from the present study, except as a possible pool for recruitment into ONR headquarters. References to ONR in this study are to ONR headquarters only.

For purposes of this report, "diversity" is defined as the presence of a significant number of women, members of underrepresented racial groups, and persons with disabilities throughout the organization. "Significant" presence is determined by the numbers of individuals in each of the above groups eligible for employment in ONR positions. For ease of reference, the phrases "minorities and women," "underrepresented groups," and "target group(s)" used in this report all refer to the three groups above, unless otherwise indicated.

The racial and ethnic groups targeted here are the primary groups that continue to be underrepresented in science and engineering fields: U.S. citizens who are African Americans, Hispanics, and American Indians. Asian Americans are not included since in most fields of science and engineering they are not underrepresented. Some issues related to Asian Americans did surface in the interviews with ONR employees, however, and it was clear that their experiences were not the same as whites. The committee recognizes that this categorization does not capture the complexity of the situation. Not all Hispanic subgroups, for example, are underrepresented, while some Asian Americans, like Pacific Islanders, are. Unfortunately, the data sources available to us do not permit a greater level of detail.

The definition of "persons with disabilities" is also complex. For purposes of this report, the definition used by NSF in its national surveys is applied: "individuals who have severe difficulty seeing, hearing, walking and/or lifting or are unable to perform these tasks." While the national surveys do collect data on these individuals, the numbers in any given field are very small.

In describing ONR's current science and engineering work force, the committee has not been able to provide any detailed analysis by race or disability status as there is only one member of a minority group (an African American) in the population of 150, and only three persons with disabilities, according to ONR records. The bulk of the analysis, therefore, focuses on differences by gender. Data tables on the ONR work force are provided in the appendixes wherever possible, but in some cases the data have been omitted or have been presented in aggregate form to avoid the possibility of associating responses with specific individuals.

ACKNOWLEDGMENTS

In carrying out this project, the committee has enjoyed sponsorship of both the Naval Studies Board and the Committee for Women in Science and Engineering, located in the NRC Office of Scientific and Engineering Personnel (OSEP). The committee also had significant interaction with Rear Admiral Mark Pelaez, former Chief of Naval Research, who sponsored the study; Dr. Fred Saalfeld, Deputy Chief of Naval Research; and the ONR Diversity Committee through Dr. Constance Oliver, its chair. ONR's Office of Human Resources, under the direction of Mary Aylor, was very cooperative in supplying personnel data. In addition, Debra Hughes, Deputy Director of the Corporate Programs Division, provided extensive background information on ONR's educational programs.

The committee is grateful for the hard work and sound advice of Ronald Taylor, Director of the Naval Studies Board, and Marilyn Baker, Associate Executive Director of OSEP, who both entered this effort after it was under way and were instrumental in its completion. It also wishes to thank Karen Bogart, who conducted extensive interviews with ONR personnel; Molla Teclemariam, who provided data analysis and technical support; and Tamae Wong, who collected examples of successful diversity initiatives in other organizations. Pamela Lohof provided administrative support for the project and did an excellent job of editing the final document.

In our world today, it is not enough for us and our students to acknowledge, in an abstract sense, that other kinds of people, with other modes of thought and feeling and action, exist somewhere—unseen, unheard, unvisited, and unknown. We must, in addition, extend ourselves in order to have direct contact with some substantial portion of that larger universe. There must be opportunities to hear different views directly—face to face—from people who believe them and embody them. Much can be learned from reading, from travel, and from formal academic study. But little if anything can substitute for the experience of continued association with others who are different from ourselves, and who challenge us—even as we challenge them.

Neil L. Rudenstine
President, Harvard University
1996

CONTENTS

PART TWO:
USING ONR'S CORPORATE PROGRAMS TO ENHANCE DIVERSITY

FIGURES

TABLES

EXECUTIVE SUMMARY

In late 1994, the Office of Naval Research (ONR) asked the National Research Council (NRC) to provide advice on how to ensure diversity in its future science and engineering work force in order to meet the needs of anticipated naval science and engineering specialties. Discussions led to the creation of a two-part study, culminating in this report.

Part One of the report examines the characteristics of the science and engineering (hereafter referred to as S&E) work force at ONR headquarters and in the national pools of experienced and recent S&E graduates. It also examines recruitment and hiring practices at ONR and the current work environment for scientists and engineers. It identifies short-term activities that the Navy can undertake to address recruitment, retention and attrition, utilization, and career development of the female, minority, and disabled scientists and engineers in ONR's work force.

Part Two of the report examines ONR's educational programs (its "corporate programs") and recommends ways they can be strengthened to enhance their overall effectiveness. It also identifies ways these programs can help increase the diversity of the nation's overall pool of scientists and engineers working in areas of relevance to the Navy and, where possible, the diversity of scientists and engineers working at ONR.

The racial and ethnic groups targeted here are the primary groups that continue to be underrepresented in science and engineering fields: U.S. citizens who are African Americans, Hispanics, and American Indians. Because there is only one member of a minority group (an African American) in ONR's S&E work force and only three persons with disabilities, the bulk of this analysis focuses on differences by gender.

PART ONE: INCREASING DIVERSITY IN ONR'S WORK FORCE

The Office of Naval Research has for 50 years been in the forefront of research and development in this country, especially in the physical sciences and engineering. In some regards, however, ONR's status may be in jeopardy. As ONR leadership recognized in requesting this study, the agency has not yet created a diverse work force. In order to remain the premier R&D agency it has been, ONR must meet the challenge of diversity with the same commitment and determination with which it pursues new research endeavors.

The National Pool of Scientists and Engineers

The committee examined national data on both experienced and recently graduated scientists and engineers in order to determine the number and characteristics of individuals most likely to be eligible for positions at ONR. The pool of experienced S&Es included U.S. citizens who received a Ph.D. between 1960 and 1989 in one of the following: biological sciences, engineering, mathematics and computer sciences, physical sciences, and relevant fields within psychology, along with recipients of master's degrees in engineering during the same time period.

Overall, there are 40,400 Ph.D.s and 23,800 master's recipients who are women, underrepresented minorities, or disabled persons in the national pools of experienced scientists and engineers. Women represent almost 15 percent of the Ph.D.s, ranging from 24 percent in the biological sciences to 9 percent in the physical sciences to 3 percent in engineering. Experienced African American, American Indian, and Hispanic Ph.D.s total 7,300 or about 3 percent. The percentage of Ph.D.s in science and engineering who are disabled is less than 1 percent.

Among experienced engineers with master's degrees, women constitute 6 percent. Underrepresented minorities are 2 percent of the population, and persons with disabilities less than 1 percent.

The data on degree recipients since 1990 show a much more diverse population. Women now receive 30 percent of the Ph.D.s in the fields under study here, including 22 percent in the physical sciences. They have also doubled their share of the master's in engineering to 13 percent. Minorities have increased fairly evenly across fields and now constitute nearly 5 percent of the total.

There are 18,500 female, minority, and disabled Ph.D.s who were educated between 1990 and 1995 in fields of potential interest to ONR. In addition, there are 9,300 members of these underrepresented groups who received master's degrees in engineering between 1990 and 1993. Combined with the population of experienced scientists and engineers, the committee believes ONR has an increasingly rich resource from which to draw in the future for its program officer and senior manager positions.

ONR's Current Science and Engineering Work Force

The scientific and engineering work force at the headquarters of ONR consists of 150 individuals. This work force includes 19 members of the Senior Executive Service (SES) and 4 GS 16 chief scientists who, for purposes of this study, will be grouped together and referred to as "senior executives." It also includes 127 program officers at the GS13, GS14, and GS15 levels. The basic demographic characteristics of the work force are relatively homogeneous. All 23 senior executives are white males. Of the program officers, 111 are male and 16 are female; 12 are Asian Americans and one is African American. Other than the one African American, there are no underrepresented minorities. Three individuals report having physical disabilities.

As would be expected, the doctoral degree dominates the educational background of the work force, especially at the higher grades. All but one senior executive hold a doctoral degree. Sixty-three percent of the male program officers and 53 percent of the female program officers in the data base also have doctoral degrees.

With 78 percent of the work force, the physical sciences and engineering clearly dominate the fields of employment, reflecting ONR's mission and its current portfolio of research and development activities. Forty-four percent of the 16 women at ONR work in one of these two areas. Overall, 75 percent of the S&E employees work at the GS15 or senior executive level; half of the women work at this level (all GS15s), with the other half primarily at the GS13 level.

The committee noted that a high percentage (79 percent) of employees listed the federal government as their most recent employer prior to coming to ONR, mostly from the Navy. By contrast, only 8 percent of the experienced Ph.D.s from underrepresented groups nationally, and 12 percent of the master's recipients in engineering, work for the federal government. Three-quarters of these individuals work in academia or industry. If ONR is to identify and attract female or minority candidates from the national pool of experienced personnel, it will have to recruit more vigorously in these other sectors. Current ONR scientists and engineers who have had some work experience outside the Navy could help in these efforts.

From 1994 to 1995, ONR hired 18 scientists and engineers: 11 men and 7 women. As a whole, both the educational level and the grade of the males were higher than that of the females. ONR is to be commended for successfully recruiting seven women in the past two years; these new hires represent almost half of the entire female S&E work force. The agency needs to be careful, however, not to concentrate new women into positions from which they are less likely to be promoted or rise to positions of leadership. The picture for minorities is not so positive; none of the 18 recent hires was a member of a minority group, and the committee could not find any evidence that ONR has yet risen to the challenge of finding and making the extra effort to recruit them.

In comparing the representation of women at ONR with those in the national pools of experienced and recent scientists and engineers, the committee concludes that ONR meets or exceeds a reasonable goal of comparability in the biological sciences, engineering, and math and computer science. Improvement is needed in the physical sciences where only 3 percent of the ONR work force is female, compared with 9 percent in the pool of experienced Ph.D.s, and in the cognitive sciences where ONR's 12 percent of women is significantly below the 34 percent in the national pool.

Assuming ONR maintains its current size, the hiring of an additional ten women would bring its overall percentage of female scientists and engineers to 17 percent, comparable to the 15 percent in the pool of experienced Ph.D.s and the 30 percent of recent doctorates. Separate efforts will need to be made to recruit or promote at least two or three women (9–13 percent) into the senior executive ranks, regardless of field.

Because of the presence of only one underrepresented minority at ONR, little can be said about the comparison of the ONR work force with the national pools. Across all fields, a minority population at ONR of five to six scientists and engineers would approach the 3–4 percent in the overall pools.

Recruitment and Hiring

In order to understand recruitment and hiring, the committee examined current ONR practices and reviewed 13 "case files" of hiring actions completed in 1993–95. Recruitment efforts appear to be very uneven and quite dependent on the preference of the selecting official, including whether there will be recruitment at all, how open it will be, and how much advertising will be done. Positions that were advertised only in the Washington, D.C. area, or restricted to individuals already in the Navy, had little chance of attracting a diverse pool of applicants, especially for senior positions.

The committee could find little evidence of concerted efforts to bring women or minorities into the recruitment pools. Advertising was generally limited to a few standard venues, some of which are mandated. Of the nine cases where the job vacancy was advertised and the number of applications is known, the total number of applications for many of these positions was surprisingly low: 210 total applications, or about 23 per position. No minorities or persons with disabilities were hired for any of the vacant ONR positions, although no data were available on how many were in the pool of applicants.

In their individual interviews, male program officers and senior executives expressed a number of opinions about the recruitment and hiring of women and minorities. Many stated that ONR was "bending over backwards" to promote diversity, but that there was an inadequate supply of minority and female applicants nationally. Others expressed concern that hiring more minorities and women would dilute the strength of the ONR work force and lower quality. Some stated that ONR could not compete successfully with industry or academia for minority and women "stars." Others, however, believed that diversity will strengthen ONR and that more aggressive measures need to be taken.

Female program officers generally believed that an adequate effort to recruit outside the organization was rarely made. Others observed that women do not apply for senior-level jobs at ONR because they believe the positions are "hard-wired" for men.

Although the number of cases is small, the picture drawn by recent hiring actions is not consistent with the view expressed by some senior executives and program officers that ONR is "bending over backwards" to promote diversity. ONR has made progress in recruiting a more diverse work force, but the committee believes there are many resources that have not yet been tapped.

The Work Environment

The committee's understanding of the work environment for ONR scientists and engineers is based primarily on 71 interviews conducted with employees by an outside consultant. In spite of long hours and the perception of inadequate staff support, most ONR employees interviewed reported that this is the best job they ever had.

In general, the women were less satisfied than the men, except in the area of income. Many women believed they are treated as second-class citizens, and some saw the work environment as hostile. White males generally saw the environment as supportive. Most saw no difference in the treatment given to women and minorities from that given to white males, and many believed that ONR was doing what it could to increase diversity. Whether fact or perception, such significant differences in employees' views of the same organization are counterproductive and undermine the ability of the agency to function as an integrated unit.

A number of program officers, especially women, described an atmosphere in job interviews, briefings, meetings, and competition for funds that was adversarial or confrontational. Many senior executives were emphatically committed to the adversarial approach to communication, stating that the ability to argue successfully for one's budget priorities is essential in a sometimes hostile bureaucratic environment, and that imposing this kind of hurdle on job candidates is important to finding the right people.

Whatever their mode of communication, a number of scientists and engineers do not relate to the prevailing management style and perceive that they have a difficult time at ONR. Providing a work environment that is supportive of all employees, not just those in the dominant groups, is critical to productivity.

ONR has initiated a number of activities in the past two years to increase the diversity of its S&E work force, including a comprehensive diversity plan and a standing diversity committee. Significant strides have been made, but much remains to be done.

Recommendations

- The Chief of Naval Research should assume the responsibility to develop specific, numerical targets for the hiring of women, minorities, and persons with disabilities into science and engineering positions at ONR. These targets should be based on a periodic assessment of the underutilization of qualified individuals from these groups, using data from national pools.

- The Chief of Naval Research should appoint an external committee composed of individuals who are experienced in the management of science and engineering and sensitive to the issues of diversity to assist ONR in achieving its diversity goals. Reporting to the Chief of Naval Research, the committee would meet periodically to review the targets ONR has set and to evaluate progress against those goals.

- ONR should expand its recruitment efforts and improve the hiring process to increase the likelihood that members of the target groups will learn about positions at ONR, will apply, and will be given serious consideration.

- ONR should improve the work environment to increase productivity, to enhance employee development, and to establish ONR as a place where women and minorities want to work.

PART TWO: USING ONR'S CORPORATE PROGRAMS TO ENHANCE DIVERSITY

Observations on ONR's Corporate Programs

Like most other mission agencies, the Office of Naval Research administers a substantial portfolio of multidisciplinary research and education programs that support its mission as a whole. The purpose of these corporate programs is to increase the number of engineers and scientists engaged in technological efforts of concern to national defense and, specifically, to the Navy. There are sixteen programs relevant to this discussion, comprising an annual budget of $35 million. They include high school apprenticeships, supplementation to research grants, graduate fellowships, support for postdoctoral researchers and young investigators, and awards to bolster science and engineering at the Historically Black Colleges and Universities (HBCUs) and Minority Institutions (MIs).

The committee was generally impressed with the breadth of educational programs ONR administers and with the dedication of its program staff. While these programs do not directly address ONR's primary mission of research and development, they are important to achieving that mission by helping to develop the scientists and engineers who will solve the Navy's technical challenges in the future.

ONR's corporate programs are relatively young, typically beginning in the early 1980s as fellowship programs. The HBCU/MI programs, especially, are only a few years old. The committee's impression is that these programs evolved historically as the Navy perceived a need, or as Congress mandated a particular activity, and while each is worthwhile in its own right, they do not necessarily form as coherent a set of programs as they might. Some of the programs are extremely small; some are restricted to specific geographic areas like the Washington, D.C. metropolitan area, which is close to ONR headquarters and several Navy facilities. Others are limited to the HBCUs and a smaller number of MIs. They need to be related in a coherent structure to a single set of goals and should reach science and engineering talent nationally.

Recommendations

- ONR should create a single, coherent vision for the corporate programs, tied more closely to the mission of ONR.
- ONR should realign its corporate programs to provide a continuum of educational opportunities from high school through postdoctoral study.
- ONR should use broader criteria for recruitment and selection.
- ONR should extend support for minority students and faculty beyond the Historically Black Colleges and Universities (HBCUs) and Minority Institutions (MIs).
- ONR should collect systematic data on program participants to allow for ongoing program evaluation.
- ONR should use the programs for postdoctoral researchers to recruit more aggressively for potential ONR employment.

PART ONE

INCREASING DIVERSITY IN ONR'S WORK FORCE

1

INTRODUCTION

The Office of Naval Research has for 50 years been in the forefront of research and development in this country, especially in the physical sciences and engineering. Predating the National Science Foundation, ONR is one of the oldest federal agencies whose mission is to fund external research and development in support of national security needs. It continues today to be viewed as a premier R&D organization where the best science and engineering is identified, supported, and applied to meet national security requirements.

In some regards, however, ONR's premier status may be in jeopardy. ONR, as a highly visible and powerful organization, is not drawing on the full range of scientific and technical talent available to administer its programs. As ONR leadership recognized in requesting this study, the agency has not yet created a diverse work force.

ONR is not alone; many R&D organizations and federal agencies are struggling with this challenge. Private corporations especially have long understood that the effectiveness and success of any organization is only as good as its people, and that to become or remain a premier organization requires recruiting and retaining the very best people from a wide range of backgrounds. No one ethnic group or gender has a monopoly on scientific talent or potential. Thus, perpetuating homogeneity is intrinsically limiting. An R&D work force that, consciously or unconsciously, excludes entire sectors of the population cannot be competitive with one that embraces the best talents from all groups. In order to remain the premier R&D agency it has been, ONR must meet the challenge of diversity with the same commitment and determination with which it pursues new research endeavors.

In building a diverse organization, ONR is far from being at a disadvantage. On the contrary, ONR is uniquely positioned to access highly qualified individuals throughout the entire U.S. research and development establishment. Through its broad responsibility of maintaining a window on R&D wherever it occurs, ONR has a potential knowledge base of candidate personnel that far exceeds most organizations both in the public and private sectors. ONR's access to universities and industry through an established network of over 5,000 principal investigators, and its ready access to R&D agencies of the government, provide it with means to identify the most capable female and minority scientists and engineers ranging from graduate students to established professionals at all levels. ONR to date has not capitalized on the potential of this intrinsic resource in evolving to a diverse work force.

THE NEED FOR DIVERSITY

The rapidly changing human resources environment in the United States today portends significant changes in the talent pool from which ONR will develop its future work force. In particular, women and ethnic minorities are a growing component of the professional scientific community. It is, therefore, incumbent on ONR to establish effective mechanisms to recruit and retain personnel from these groups and to develop a climate that is conducive to integrating diverse employees into a more productive organization. In requesting a study, the leadership at ONR has recognized this need.

But the intrinsic value of diversity to the vitality of a modern enterprise goes well beyond strictly demographic considerations. The confluence of disparate points of view, the clash of ideas, and the multiplicity of perspectives born of distinct cultural backgrounds, if managed well, can yield higher creativity and lead to greater innovation, both of which are critical to the conduct of science.

Just as important as the utility of diversity is the need, in the committee's view, to overcome traditional, institutionalized practices that, whether intentional or not, serve to exclude individuals on the basis of ethnicity or gender. This country aspires to the principle of equality of opportunity, not only because of its inherent fairness, but also because the pursuit of that principle helps to ensure that the widest possible range of human talent is available for the nation's needs.

The changing American demographics are well known. According to the National Science Foundation (1994), "[p]rofound changes are likely to occur in the composition of the U.S. population over the next half-century" (p. 11). Assuming population growth were to continue at the current projected rate, researchers at the Bureau of the Census estimate that

> the total non-Hispanic white proportion of the U.S. population would decline from 76 percent in 1990 to 53 percent in 2050. . . . [and] by 2012 more blacks than non-Hispanic whites would be added to the population each year. In 2030, the non-Hispanic white population would be less than half of the under-18 population . . . Over the following 20 years, American Indians, Asians and Pacific Islanders, Hispanics of all races, and blacks would together far outnumber the total white non-Hispanic population of elementary school children, high school students, and new entrants into college, the work force, and the military (NSF 1994, 11).

African Americans, Hispanics, and American Indians already comprise almost 40 percent of new births, one-quarter of the overall labor force, 28 percent of the college-age population, and nearly 5 percent of new Ph.D.s in the sciences and engineering. Women constitute more than 42 percent of the general full-time labor force, more than half of the undergraduate enrollment, and 30 percent of the new science and engineering Ph.D.s. From the point of view of demographics, diversity is a fact and should be a defining factor in the work force of any particular enterprise today.

In addition, diversity offers practical benefits to employers and other organizations. Several studies have shown that, based on measures of performance such as creativity, quality of problem-solving, and effective decision-making, heterogeneous groups, if managed effectively, have significantly higher potential than homogeneous groups. In a brief but comprehensive review of research on the subject, Taylor Cox and Stacy Blake (1991) summarize the findings from a number of such studies.

Rosabeth Moss Kanter (1983), for example, found that innovative companies consciously form heterogeneous teams which generate a broad spectrum of ideas and a range of approaches to problems, thus increasing the probability of novel solutions:

> Note that it is not just *any* team that aids innovation but a tradition of drawing members from a diversity of sources, a variety of areas. Innovating companies seem to deliberately create a "market-place of ideas," recognizing that a multiplicity of points of view need to be brought to bear on a problem. It is not the "caution of committees" that is sought—reducing risk by spreading responsibility—but the better idea that comes from a clash and an integration of perspectives (p. 167).

Studies at the University of Michigan by Richard Hoffman and Norman Maier (1961) quantified the higher probability of superior problem solving from groups with diverse personality characteristics and gender. They found that "[h]eterogeneous groups produced a higher proportion of high quality solutions than did homogeneous groups to three of the four problems with quality components," and that "Mixed-Sex groups tended to produce higher quality solutions than did All-Male groups" (p. 407).

Susan Jackson (1989) has also summarized research on the greater contribution of groups with diverse "personal attributes" to creative idea-generation and decision-making:

> Several reviews covering research on this topic have reached the conclusion that heterogeneous groups are more likely than homogeneous groups to be creative and to reach high-quality decisions. . . . This finding holds for a variety of personal attributes, including personality . . . types of training . . . and attitudes.
>
> Studies of research scientists similarly have shown that groups with fluid membership are likely to be more creative . . . even when groups are initially interdisciplinary. When scientists or interdisciplinary teams worked closely together on a daily basis, within three years they were found to have become homogeneous in their perspectives and approach to solving problems (pp. 148–49).

Examining diversity of opinion rather than personal characteristics, Charlan Nemeth (1986) conducted controlled experiments to test the contributions of majority and minority influence on group decision-making. In this context, "majority" refers to those in the group holding the prevailing view and "minority" to those holding a different view. Nemeth found that

> majorities foster convergence of attention, thought, and the number of alternatives considered. Minority viewpoints are important, not because they tend to prevail but because they stimulate divergent attention and thought. As a result, even when they are wrong they contribute to the detection of novel solutions and decisions that, on balance, are qualitatively better. The implications of this are considerable for creativity, problem solving, and decision making, both at the individual and group levels (p. 23).

Nancy Adler (1986) reports on studies of cross-cultural groups at the University of California at Los Angeles, which demonstrated that "[m]ulticultural teams have the potential to become the most effective and productive teams in an organization" because their members contribute "more and better ideas." Such groups are "less likely to subconsciously limit their perspectives, ideas, conclusions, and decisions to that of the majority or group leadership" (pp. 110–11). At the same time, diverse groups have the potential to become the least productive if not managed effectively. Mistrust, stereotyping, and poor communication can impede progress toward common goals. Adler goes on to talk about how to manage cultural diversity for maximum effectiveness.

Cox and Blake (1991) also point out that diversity generates greater organizational flexibility: cultures and structures that are less standardized, more fluid, more agile, more adaptable. Organizations with these attributes, they point out, are essential in today's world and can react much more quickly to changes in the environment.

Examples of the value of diversity in science and engineering abound. The influx of women into biomedical research has contributed substantially to understanding sexual differences in a wide range of diseases and health-related pathologies. The presence of women in anthropology, bringing new approaches to the study of social behavior in primates, led to major revisions in primatology. In physics, feminist thinking is making a contribution to the reassessment of competing philosophies, traditions, and priorities (Whitten 1996).

In addition to internal benefits to an organization, diversity brings external ones. As non-whites increase, they will make up a growing fraction of the voting public. Consequently, successful results in congressional appropriations, the committee believes, will increasingly require credibility with legislators, not only on the basis of quality, quantity, and societal value of output, but also with respect to hiring practices, equal opportunity, and access to jobs at the agencies seeking funding. Those agencies perceived as closed, discriminatory, not reflective of the taxpayer and voter population, or not earnest in pursuing diversity will have an added burden in maintaining their funding.

Mirroring the diversity of one's clients or customers, the committee believes, reaps benefits for governmental and commercial organizations alike. Ambassador Andrew Young, co-chairman of the 1996 Atlanta Olympic Committee, offers an example of the political currency of diversity in the international arena. He compares the experience of his Atlanta committee with that of the Miami World

Cup Committee. Miami, one of the most diverse cities in America and an initial favorite for the World Cup championship, formed a committee that was a subgroup of the Chamber of Commerce. All members were white males. The decision-makers on the International World Cup Committee were appalled at the lack of representation from the diverse population groups of Miami, and this is widely believed to have influenced the committee's decision to go elsewhere. Two-thirds of the Atlanta Olympic Committee, on the other hand, were women. Also represented were Asian Americans, African Americans, Hispanics, and others—virtually every constituent of the Atlanta population. The 1996 Olympic Games brought substantial new revenue as well as immeasurable pride to the Atlanta community.

Diversity is also important in recruiting new employees. Recruiting top scientific and engineering talent is fiercely competitive, and the attributes of the existing work force are of paramount importance in establishing a competitive position in hiring. The strongest candidates will be anxious to join a team of the highest caliber. Given that diversity is a critical ingredient in work force quality, an agency with a homogeneous work force, or one perceived as discriminatory, will find recruiting to be increasingly difficult. Even those white males with the highest qualifications may be reluctant to come to an artificially homogeneous organization which limits their opportunities to work on diverse teams and to acquire the skills that other employers require, including the experience necessary to become effective managers of a diverse population.

THE CURRENT ENVIRONMENT

The committee's consideration of diversity in the S&E work force at ONR must be placed in the context of the changing political and economic environment. The emphasis on working toward a balanced budget, together with the removal of the USSR as a military threat to the United States, have led to a significant downsizing of the nation's military forces. In this atmosphere, programs of every kind come under repeated scrutiny to see if they are necessary, and long-term investment frequently falls prey to immediate needs. Neither the Navy nor ONR has been immune to these forces.

In addition, ONR has recently undergone a major reorganization. The Navy previously segregated basic research (6.1), applied research (6.2), and exploratory development (6.3) in separate organizational units: the Office of Naval Research (ONR), the Office of Naval Technology (ONT), and the Office of Applied Technology (OAT). In 1994 these three units were joined into a single organization in which program officers manage an appropriately related mix of 6.1, 6.2, and 6.3 funds. An intent of the merger was to relate the objectives of 6.1 research much more closely to perceived developmental requirements of the Navy expressed through 6.2 and 6.3 programs. It has had a separate effect, however, of bringing together individuals with disparate educational and work backgrounds who earlier managed either basic or applied research, but not both. The complexities raised by this reorganization provide an important overlay on the environment at ONR.

The reduction in the size of military forces and the reorganization of ONR have also led to painful reductions in grade of a number of individuals at the senior executive level. The resultant personnel freezes, lack of growth, and uncertainty of the future tend not only to reduce turnover, with the consequent significant decrease in the opportunity to make changes in the composition of the entire organization, but also to increase the highly competitive struggle for diminishing program funds.

Despite these current circumstances —or perhaps because of them—the committee believes that increasing the diversity of ONR's work force is essential. An organization with fewer staff with more far-reaching responsibilities is especially in need of finding and keeping the best talent to ensure that it is operating efficiently and funding the most promising research and development efforts. Finding the best talent implies searching in the complete pool of qualified individuals of whatever gender or ethnicity and not limiting that search to the traditional white male population. The committee expects that both the quality and the diversity of the ONR work force will be increased through greater efforts in this direction.

In addition, creating a healthy work climate will be equally important. Environments perceived as hostile or unsatisfying for particular groups have very real, quantifiable costs in personnel turnover, retraining, absenteeism, and lost productivity (Cox and Blake 1991). Training in making effective use of diversity will, therefore, be essential. Both managers and employees must be able to understand differences, build group cohesiveness, establish shared values, communicate across cultural and gender boundaries, and resolve conflicts. If ONR can inculcate these skills in a work force that truly reflects the increasing diversity of the nation's S&E work force, it will be well positioned to meet future challenges.

2

PROFILES OF THE SCIENCE AND ENGINEERING WORK FORCE

THE NATIONAL POOL OF SCIENTISTS AND ENGINEERS

Before addressing the diversity of the ONR work force, it is important to understand the characteristics of the overall national pool of scientists and engineers from which those individuals are drawn. Based on data on current ONR employees and ONR's description of its expectations for hires, the committee identified a pool of U.S. scientists and engineers which it believes contains those individuals most likely to be eligible for ONR positions at the level of program officer or higher. This pool is limited to individuals with either a doctoral degree in a relevant science or engineering field or a master's degree in engineering. It is further limited to those with U.S. citizenship, a requirement of employment at ONR. The committee examined the characteristics of the overall pool and, to the extent possible, the corresponding characteristics of the members of underrepresented groups. None of these characteristics indicate an eligibility for or interest in work at ONR *per se*, but they do point to the types of individuals who have the potential for a good match. With the exception of the data on recent Ph.D.s, these numbers are estimates based on surveys of a representative sample of Ph.D.s and master's recipients currently working in the U.S.

Because most scientists and engineers have several years of professional experience before joining ONR, the committee focused its attention on the pool of individuals who received their degrees between 1960 and 1989. Deemed to be of an experience level comparable to those currently at ONR, this group is described in the section below entitled "Experienced Doctorate and Master's Recipients." However, more recent degree recipients are by now or soon will be of an experience level to make them eligible for ONR positions as well. The characteristics of this pool are described in the section entitled "Recent Doctorate and Master's Recipients."

The pools of scientists and engineers described here cover all sectors of employment, including academia, industry, and government, and all geographic regions of the U.S. This is the case even though, as the report discusses later, some ONR recruitment efforts are restricted to the

Washington, D.C. metropolitan area, and many ONR employees are hired from elsewhere in the Navy. In addition, the committee recognizes that job mobility across geographic regions or across sectors of employment may be limited for many scientists and engineers.

At the same time, though, the data do not permit identifying a pool of potential employees by field for a specific geographic region or just for the Navy. Also, it is impossible to tell which individuals from other sectors of employment should be excluded because they are not truly mobile. Finally, while acknowledging ONR's unique job requirements, the committee believes that drawing from a broad national pool whenever possible can only enhance the quality and diversity of potential job candidates.

Experienced Doctorate and Master's Recipients

Ph.D.s in Science and Engineering Fields

In order to identify the pool of underrepresented minorities, women, and persons with disabilities possibly eligible to assume current management positions at ONR, the committee examined data from the 1993 Survey of Doctorate Recipients (the most recent data were not available at the time of committee deliberations). In order to match the pool to ONR's needs, only individuals with Ph.D.s in the following fields were included: biological sciences, engineering, mathematics and computer science, physical sciences, and relevant fields within psychology. Only U.S. citizens were included, and only those who indicated that they were currently working in a field "somewhat related" or "closely related" to their Ph.D. field (91 percent of the total). In addition, only individuals who received their doctorates between 1960 and 1989 were included. Data on this pool are presented in Table Series A-1.

The overall pool of experienced Ph.D.s who meet the above criteria includes 228,400 individuals. Women represent almost 15 percent (33,600) of the total pool of experienced Ph.D.s, ranging from 24 percent (16,200) in the biological sciences, to 9 percent (5,900) in the physical sciences, to 3 percent (1,700) in engineering (see Figure 2–1). The percentage of women is highest in psychology (34 percent or 7,600), but this field represents only 4 percent of the expertise of ONR's current work force.

There are 7,300 African Americans, American Indians, and Hispanics in this pool, comprising about 3 percent of the total. Like women, underrepresented minorities are represented in somewhat higher proportions in psychology. Asian Americans and Pacific Islanders comprise 9 percent (21,000) of the overall pool, with the highest concentration in engineering. Disabled persons represent less than 1 percent (1,500).

Within the physical sciences, the greatest number of women are in chemistry (11 percent or 4,000) and the lowest number in atmospheric sciences (2 percent or 30) and physics (4 percent or 800), as depicted in Figure 2–2. Although the numbers of underrepresented minorities in some of these fields are extremely small, they are somewhat higher in chemistry (3 percent or 1,200) and physics (2 percent or 500).

There has been substantial growth in the number of female Ph.D.s over the past few decades. Among graduates who received their degrees in 1960–64, women

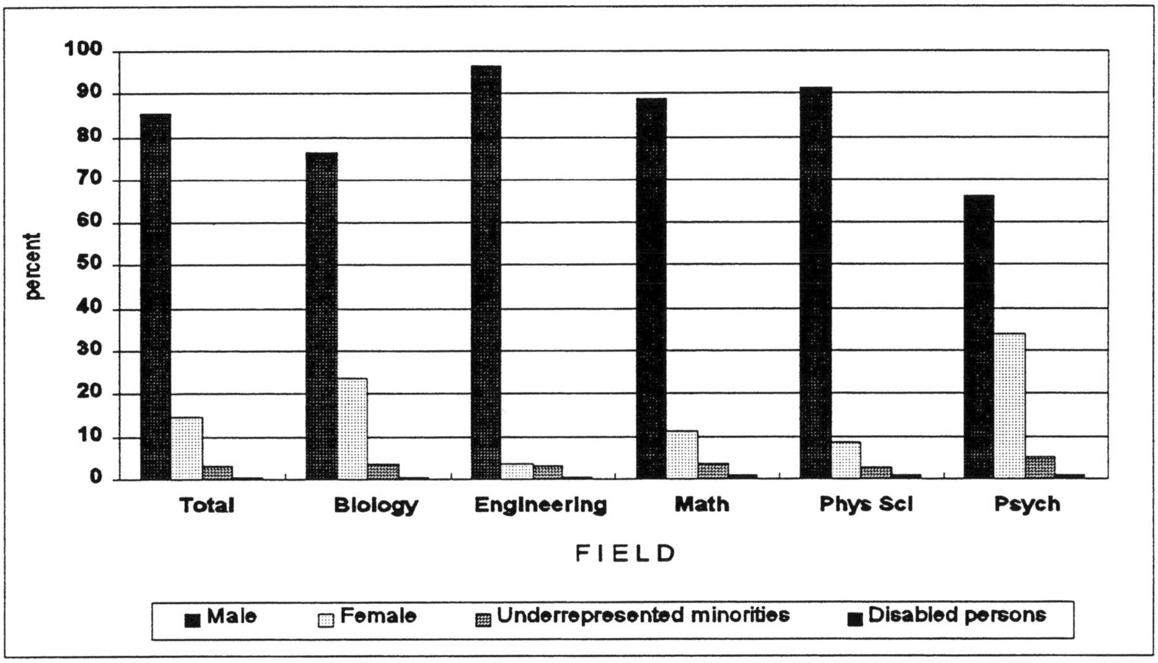

Source: Table A-1.1

FIGURE 2–1 Experienced S&E doctorates by field, gender, race, and disability status.

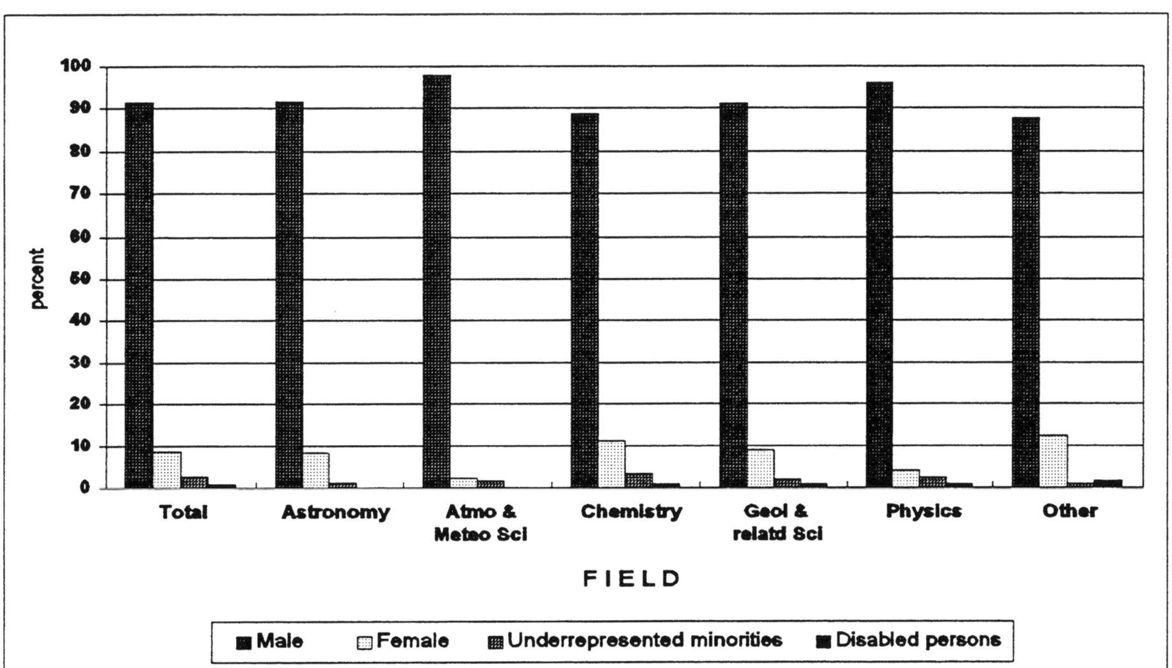

Source: Table A-1.2

FIGURE 2–2 Physical science doctorates by field, gender, race, and disability status.

represented less than 5 percent (600), but they accounted for 25 percent (11,000) of 1985–89 graduates (see Figure 2–3).

Except for Hispanics, Ph.D. degrees awarded to underrepresented minorities have increased only modestly over this same time period. Degrees to African Americans, American Indians, and Hispanics together increased from about 120 in 1960–64 to 1,700 in 1985–89, an increase from 1 to nearly 4 percent of the total. Since the late 1970s, there has been virtually no increase in the number of underrepresented minorities receiving Ph.D.s in these fields.

The percentage of Ph.D.s in science and engineering who are disabled remained fairly constant at less than 1 percent between 1960 and 1989. Within that population, the numbers of individuals reporting disabilities were higher in the earlier cohorts and lower in the most recent group. When asked on the 1993 Survey of Doctorate Recipients, 340 of the 1965–69 graduates reported having a disability, compared to 170 of the 1985–89 graduates. This is not surprising, given the fact that disability increases with age, and the former group is approximately 20 years older than the latter.

Some additional information is known about the characteristics of the target population through the 1993 Survey of Doctorate Recipients. There are about 40,400 individuals from the national pool of experienced Ph.D. scientists and engineers who are underrepresented minorities, women, or disabled persons (see Table

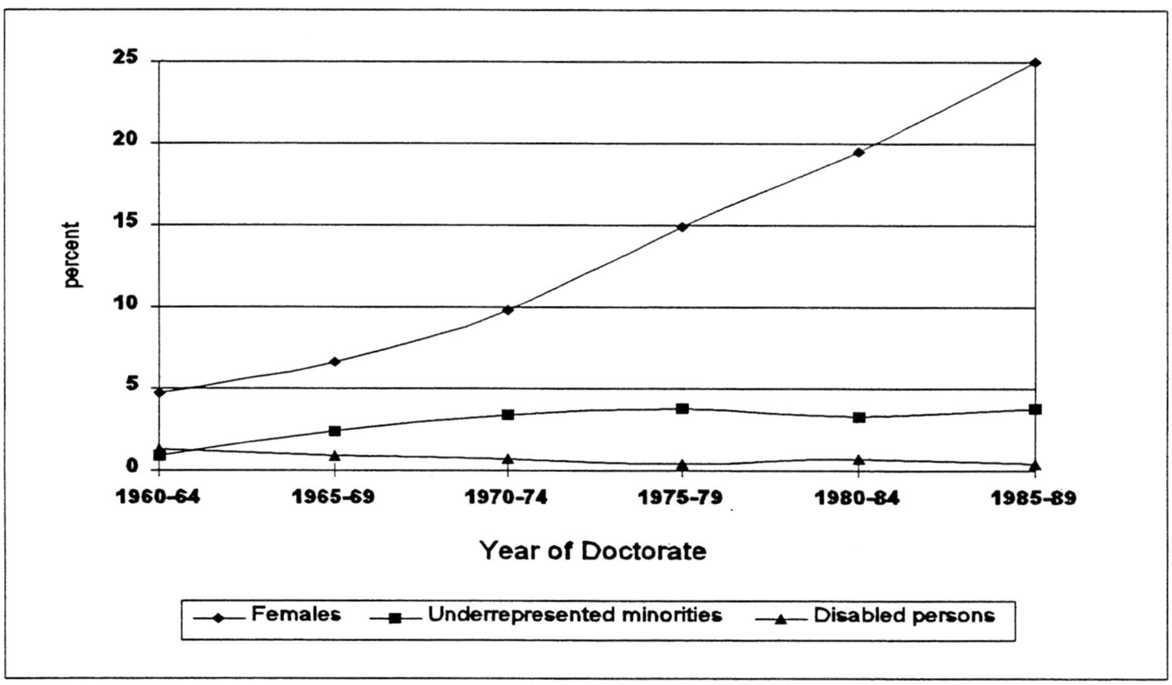

Source: Table A-1.3

FIGURE 2–3 Percent female, underrepresented minorities, and disabled S&E doctorates: 1960–89.

A-1.5). Half of this population (20,300) are employed in universities and four-year colleges, with another one-quarter (10,000) employed in private-for-profit companies (see Figure 2–4). Only 3,000 (or 8 percent) are employed in the federal government.

Of the 50 percent of the above target group employed in universities and four-year colleges, nearly half (9,400) of those are in research universities. They are divided fairly evenly across academic ranks, with about 25 percent each among full professor, associate professor, and assistant professor. The remaining 25 percent are instructors or adjunct faculty or hold non-faculty positions.

Attempting to understand what kind of work these individuals do is, of course, more difficult. However, 63 percent (25,300) of them report that they supervise others, and 79 percent (32,000) are involved in conducting research or development.

Master's in Engineering

Information on the characteristics of experienced individuals with master's degrees in engineering was drawn from the 1993 National Survey of College Graduates. As with the Ph.D.s, the pool contains only U.S. citizens working in a field "somewhat related" or "closely related" to their field of

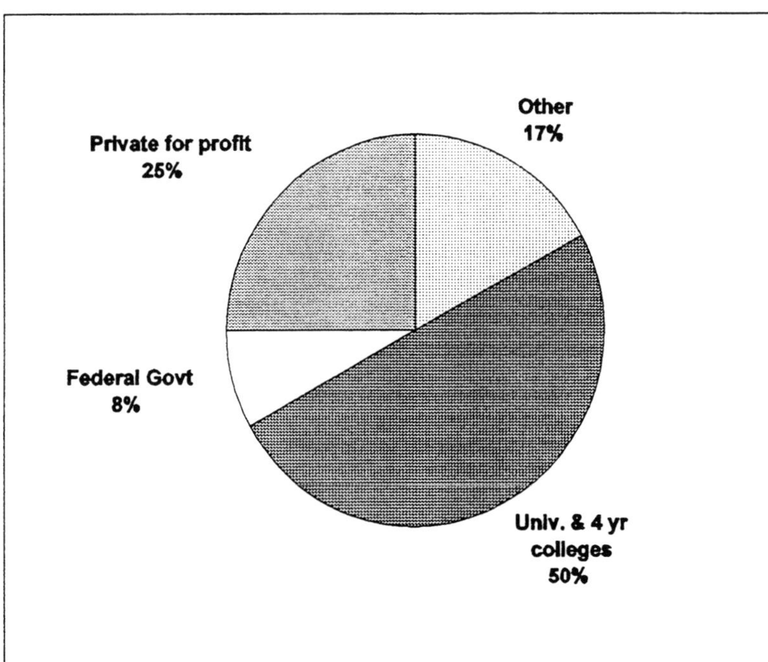

Source: Table A-1.5

FIGURE 2–4 Underrepresented S&E doctorates by sector of employment.

study (72 percent of the total) who received their master's degrees between 1960 and 1989. Data are provided for engineering as a whole and for the following five areas: electrical and computer science, mechanical and industrial, civil and architectural, chemical, and other engineering fields (see Table Series A-2).

Women represent 6 percent (17,900) of this pool, with the highest percentage of women being in chemical engineering (11 percent or 2,700) and the lowest in electrical engineering and computer science (4 percent or 3,900) (see Figure 2–5). Underrepresented minorities are 2 percent of the population, or about 5,700 individuals. Persons with disabilities constitute less than 1 percent, or about 1,300 individuals.

Since the early 1960s, women have increased their representation in engineering from 2 percent (400) of the master's degree recipients in 1960–64 to almost 13 percent (8,700) in 1985–89 (see Figure 2–6). Minorities (excluding Asian Americans) have increased from 200 to 2,000, to about 3 percent of the population. Experienced master's degree engineers who reported having a disability in 1993 average 0.5 percent or less of the pool regardless of when they received their master's degree, with the one exception of the earliest cohort of 1960–64 graduates (1.4 percent).

This population of engineers includes a pool of 23,800 experienced master's-degree holders who are members of the target groups (see Table A-2.3).

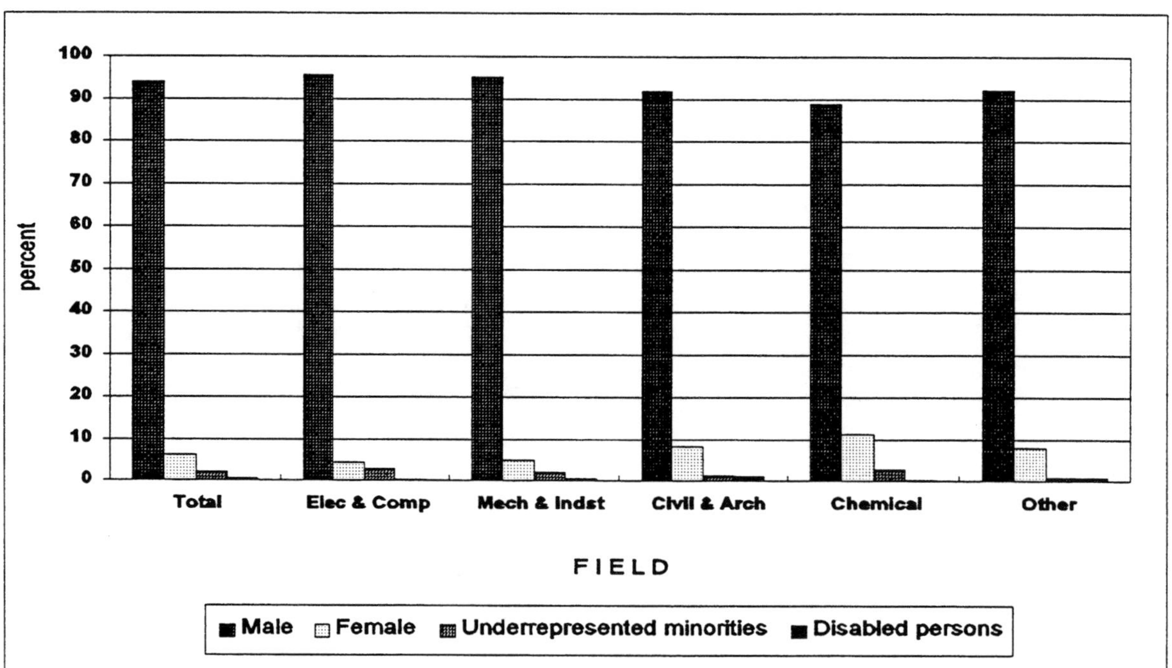

Source: Table A-2.1

FIGURE 2–5 Experienced master's in engineering by field, gender, race, and disability status.

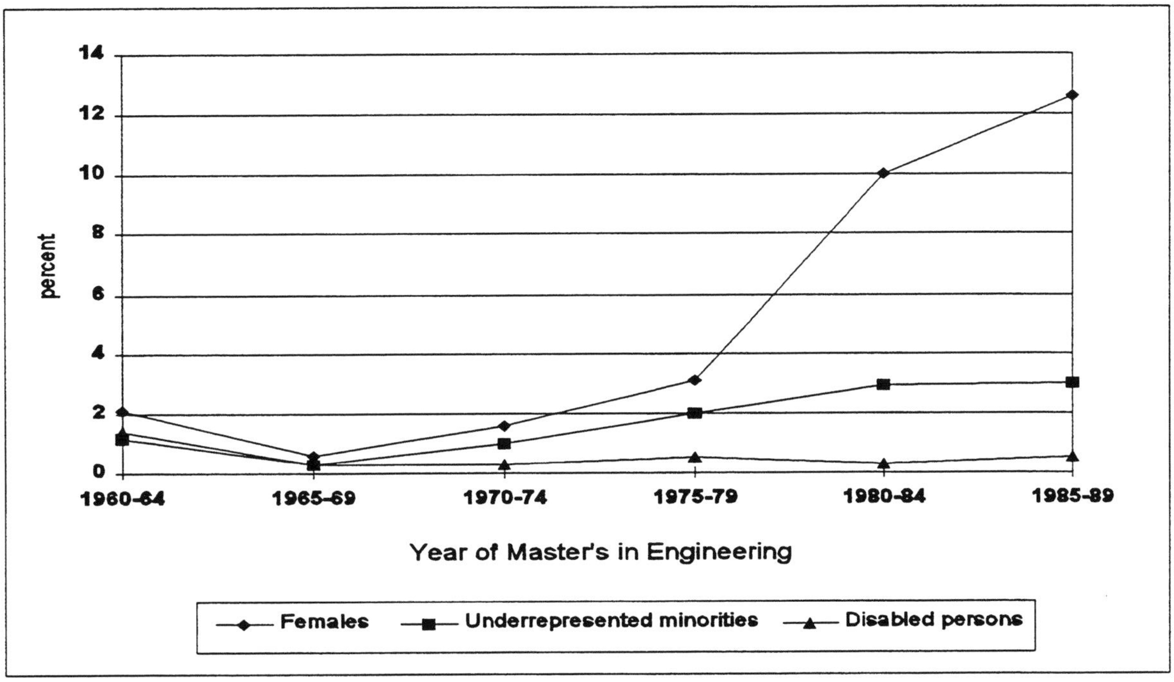

Source: Table A-2.2

FIGURE 2–6 Percent female, underrepresented minorities, and disabled persons with master's in engineering: 1960–89.

Seventy-two percent (17,100) of these engineers work for private-for-profit companies; 12 percent (3,000) work for the federal government; and only 4 percent (1,000) work for universities and four-year colleges (see Figure 2–7). As would be expected, almost two-thirds of them have over 10 years of professional experience, while the other one-third have 5 to 10 years of experience. Eleven percent (2,700) of them indicate that they are top or mid-level managers, executives, or those who manage other managers. Nearly 60 percent (14,000) of this target population are engaged in conducting research or development.

Findings and Conclusions

Based on these data, the committee finds that, even though the percentages are low, there are a substantial number of women, minorities, and disabled persons who are experienced scientists and engineers potentially eligible for employment at ONR. At the Ph.D. level, the numbers of women are higher in the biological sciences (24 percent) and lower in the physical sciences (9 percent) and engineering (3 percent). Even in physics, however, the committee estimates that there are 800 Ph.D. women with the appropriate background for possible ONR employment, and 1,600 female Ph.D.s in engineering. Among master's educated engineers, women are still a small percent (6

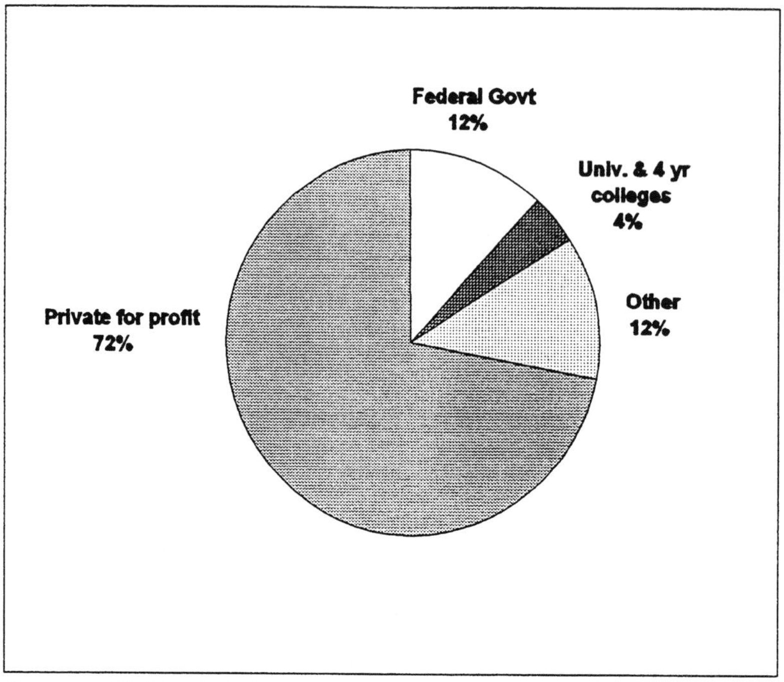

Source: Table A-2.3

FIGURE 2–7 Underrepresented master's in engineering by sector of employment.

percent), but that proportion doubles among those educated in the late 1980s. In general, the numbers of eligible women increase significantly among scientists and engineers who have completed their degrees in the past 10–15 years, providing many more potential candidates for ONR positions.

The picture for minorities is less positive; they remain at about 3 percent (7,300) of the Ph.D.s and 2 percent (5,700) of the master's engineers. Even among more recent cohorts (1985–89 graduates) they represent less than 4 percent (1,700) of the Ph.D.s and 3 percent (2,000) of the master's in engineering.

Overall, there are 40,400 Ph.D.s and 23,800 master's recipients who are women, underrepresented minorities, or disabled persons in these national pools of experienced scientists and engineers. Three-quarters of them work in academia or industry. Only 8 percent (3,000) of the Ph.D.s across all fields work for the federal government, with the highest concentration (10 percent) in engineering. Among master's recipients in engineering, only 12 percent (3,000) overall work for the federal government, with the highest concentration in civil and architectural engineering (20 percent or 1,100). If ONR is to identify and attract candidates from the national pool of experienced personnel, it will have to recruit more vigorously where most of those individuals are—in academia and industry.

Recent Doctorate and Master's Recipients

As mentioned above, ONR indicated to the committee that it did not hire new or very recent graduates for its program officer or senior executive positions. However, the pool of recent degree recipients will soon become eligible for ONR positions, and it is useful in planning a long-term strategy for increasing diversity to examine some of the characteristics of this pool of potential ONR employees.

Ph.D.s in Science and Engineering Fields

For purposes of this study, recent Ph.D.s were defined as those receiving their degrees between 1990 and 1995, using data from the annual Survey of Earned Doctorates. As above, only U.S. citizens are included. Data tables are available in Table Series A-3.

In contrast to the earlier cohorts, Ph.D. recipients in the early 1990s are more diverse (see Figure 2–8). Of this population, 30 percent (16,700) are female, ranging from 15 percent (1,600) in engineering, 22 percent each in the physical sciences (3,200) and math and computer science (1,100), to 41 percent (7,800) in the biological sciences. Over half (3,000) of the new Ph.D.s in psychology are women.

Underrepresented minorities represent about 4.5 percent (2,400) of the new Ph.D.s, spread fairly evenly across fields except for mathematics and computer science, which have 2.9 percent (135). Disabled persons represent 1.2 percent (670) of the population. This percentage cannot be compared with the percentage of disabled persons in the more experienced pool because the question is asked differently on the two surveys from which these data are drawn (see the notes to Tables A-1.1 and A-3.1). In any case, the numbers in both pools are extremely small.

Within the physical sciences, women represent 27 percent (2,000) of the chemists and 24 percent (500) of the geologists, but still only 10 percent (400) of the physicists (see Figure 2–9). Non-Asian minorities (4 percent or 600) are represented in somewhat higher numbers in chemistry (5 percent or 300), as before, and less in physics (3 percent or 100). These data revealed about the same percent (1 percent) of disabled persons in the physical sciences as in the five broad fields overall.

Nearly 18,600 Ph.D.s have been awarded in the past six years to underrepresented scientists and engineers (see Table A-3.3). As is the case with all recent Ph.D.s, two-thirds of the degrees to members of underrepresented groups came from institutions in four regions of the country: South Atlantic, Middle Atlantic, eastern North Central, and Pacific. Research universities awarded 84 percent of the Ph.D.s to underrepresented scientists and engineers, only slightly less than the 87 percent awarded to all Ph.D.s.

Master's in Engineering

Data on recent master's recipients in engineering are available from the National Survey of College Graduates for 1990–93 only; 1994 and 1995 graduates are not included. As with the Ph.D.s, U.S. citizens working in a field "somewhat related" or "closely related" to their degree were included (see Table Series A-4).

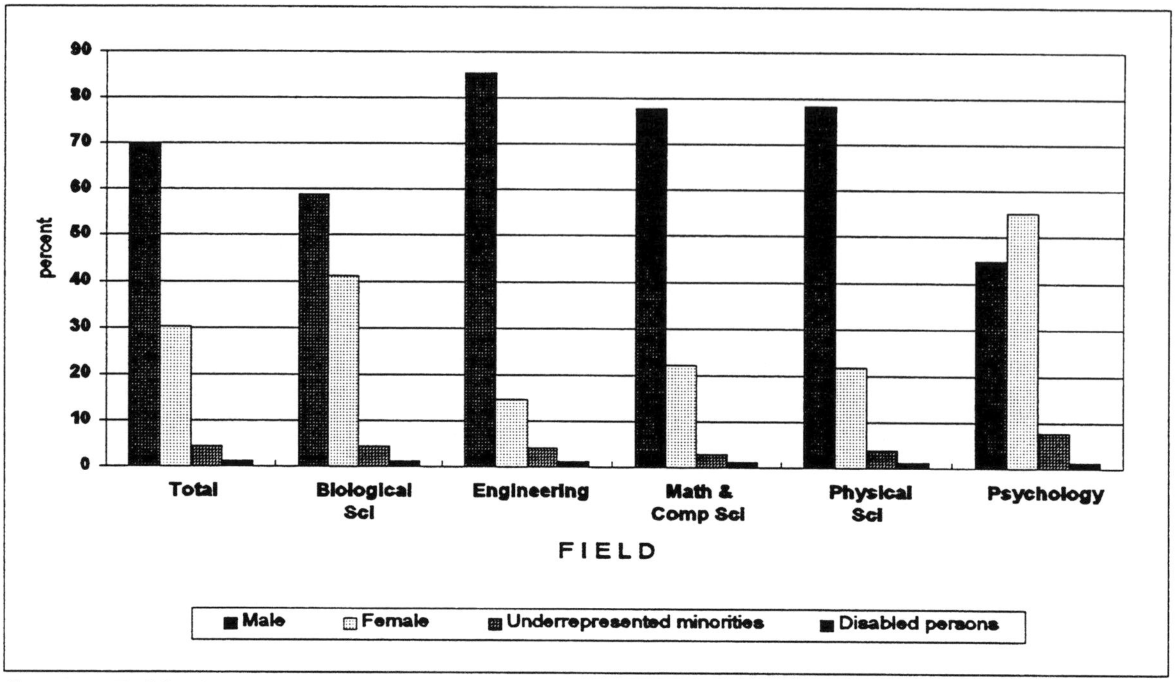

Source: Table A-3.1

FIGURE 2–8 Recent doctorates by field, gender, race, and disability status.

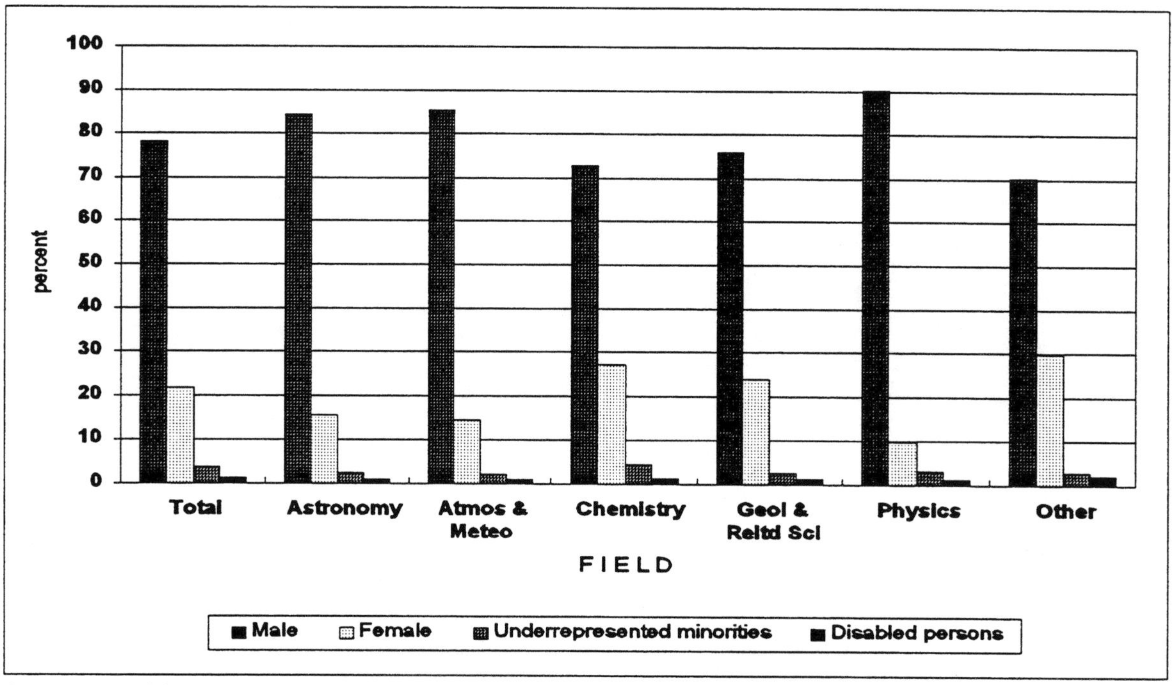

Source: Table A-3.2

FIGURE 2–9 Recent doctorates in physical sciences by field, gender, race, and disability status.

The number of women receiving master's degrees in engineering in recent years has indeed increased; women represent 13 percent (7,400) of the population in this newer cohort, while they were only 6 percent (17,900) of the more experienced group (see Figure 2–10). Underrepresented minorities overall account for 4 percent (2,400). African Americans represent 3.7 percent (2,100) of this population, compared to 1.4 percent (4,100) of the more experienced group. American Indians and Hispanics, however, appear unchanged at less than 1 percent each. The number of disabled persons in the survey sample for this population was too small to permit an estimate.

There are approximately 9,300 engineers who are in the targeted population (see Table A-4.2), educated predominantly in electrical engineering and computer science or in mechanical and industrial engineering. As with Ph.D. recipients, they received their degrees primarily from schools in the South Atlantic, Middle Atlantic, eastern North Central, and Pacific regions of the U.S.

Findings and Conclusions

The data on degree recipients since 1990 show a much more diverse population. Women now receive 30 percent of the

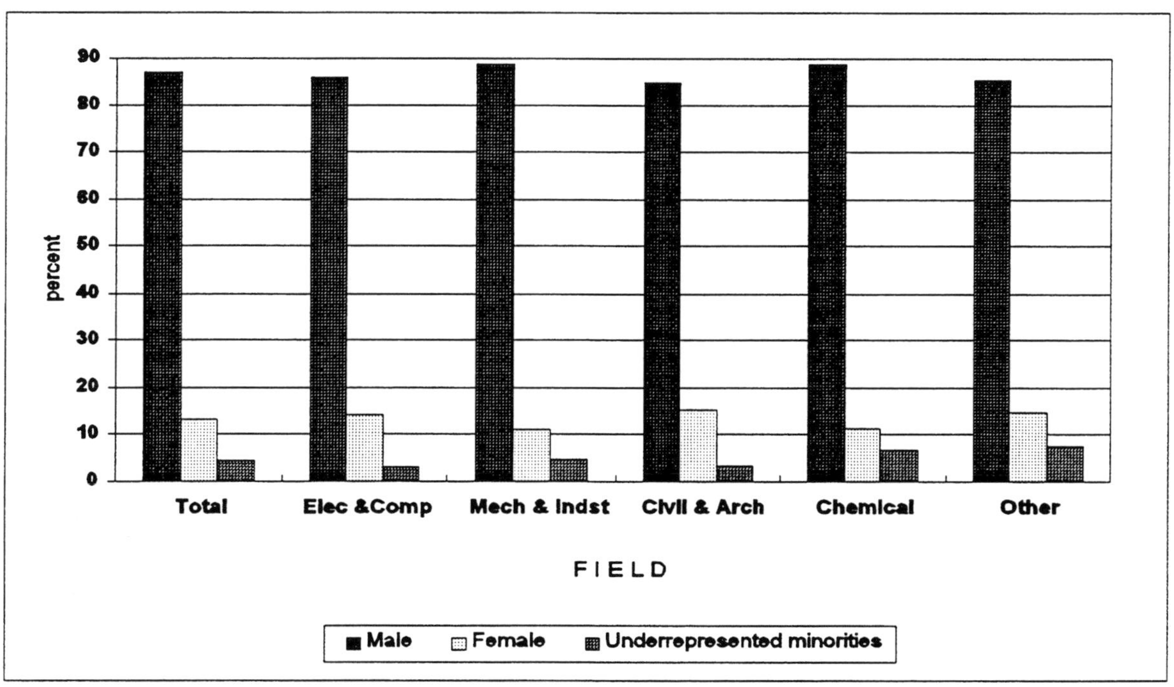

Source: Table A-4.1

FIGURE 2–10 Recent master's in engineering by field, gender, and race.

Ph.D.s in the fields under study here, including 22 percent in the physical sciences. They have also doubled their share of the master's in engineering to 13 percent.

Minorities have increased fairly evenly across fields and now constitute nearly 5 percent of the total. African Americans have made important strides among master's recipients in engineering; they represent 4 percent of recent degree recipients. But American Indians and Hispanics have not increased measurably.

There are 18,500 female, minority, and disabled Ph.D.s who were educated between 1990 and 1995 in fields of potential interest to ONR. In addition, there are 9,300 members of those underrepresented groups who received master's degrees in engineering between 1990 and 1993. Combined with the population of experienced scientists and engineers, the committee believes ONR has an increasingly rich resource from which to draw in the future for its program officer and senior manager positions.

ONR'S CURRENT SCIENCE AND ENGINEERING WORK FORCE

This section describes the demographic, educational, and employment characteristics of the 150 individuals who constitute the scientific and engineering work force at the headquarters of the Office of Naval Research. This work force includes 19 members of the Senior Executive Service (SES) and 4 GS16 chief scientists who, for purposes of this study, will be grouped together and referred to as "senior executives." It also includes 127 program officers at the GS13, GS14, and GS15 levels.

Most of the information is drawn from data on the entire ONR S&E work force of 150, compiled from ONR's Office of Human Resources data base of February 1996. These data are presented in Table Series A-5.

Demographics

The S&E work force is relatively homogeneous in its basic demographic characteristics (see Figure 2–11). All 23 senior executives are white males. Of the 127 program officers, 111 are male and 16 are female; 12 are Asian Americans, and 1 is an African American male. Other than the one African American, there are no underrepresented minorities. According to ONR records, three individuals report having physical disabilities.

As stated earlier, because Asian Americans are generally not underrepresented in the science and engineering work force, most of this discussion groups them with their white counterparts. Because of the small number of minorities and persons with disabilities, no separate analysis by race or disability status is useful.

The ONR S&E population spans more than four decades in date of entry into college, which can be used as a surrogate for age. Three-quarters of the work force entered college before 1970, putting them generally over age 45 (see Figure 2–12). As would be expected, 20 out of 23 of the senior executives entered college before 1970, and 11 entered college before 1960. Differences between the senior executives and GS15s in age are negligible; however, over half of the program officers at the GS13 and GS14 grades entered college after

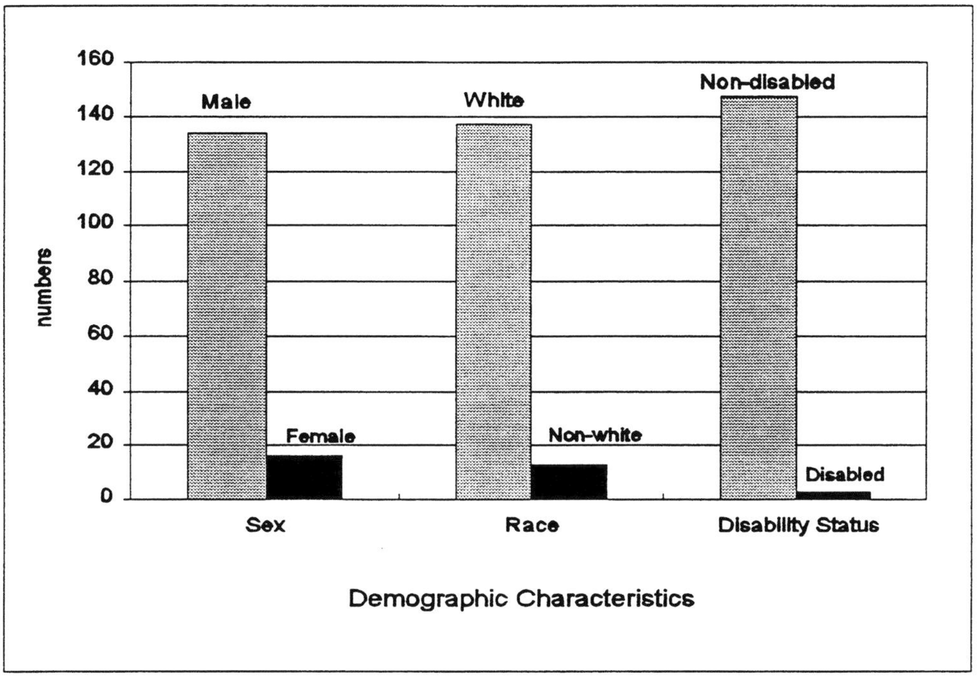

SOURCE: Table A-5.1

FIGURE 2–11 ONR S&E employees by demographic characteristics.

1970. Seven program officers did not enroll in college until after 1980.

Findings and Conclusions

The data described above present a picture of a fairly homogeneous population. Like the national pools, the majority of the 150 scientists and engineers at ONR are male, white, and age 45–65. By its sheer numbers and similar characteristics, this group is likely to establish the dominant culture at ONR. By contrast, the 16 female scientists and engineers, who represent about 10 percent of the population, are generally younger.

The absence of any women or minorities among senior executives is evident. It is understandable that most of the senior managers would be white males, given the relatively small percentage of women and minorities in their age cohort and the lack of turnover at ONR. At the same time, though, no women or minorities have been hired or promoted into these positions, even when they did become vacant. Until senior executives become more diverse, the message of diversity to the program officers will have a limited effect.

Education

The doctoral degree dominates the educational background of the work force, especially at the higher grades (see Figure 2–13). All but one senior executive hold a

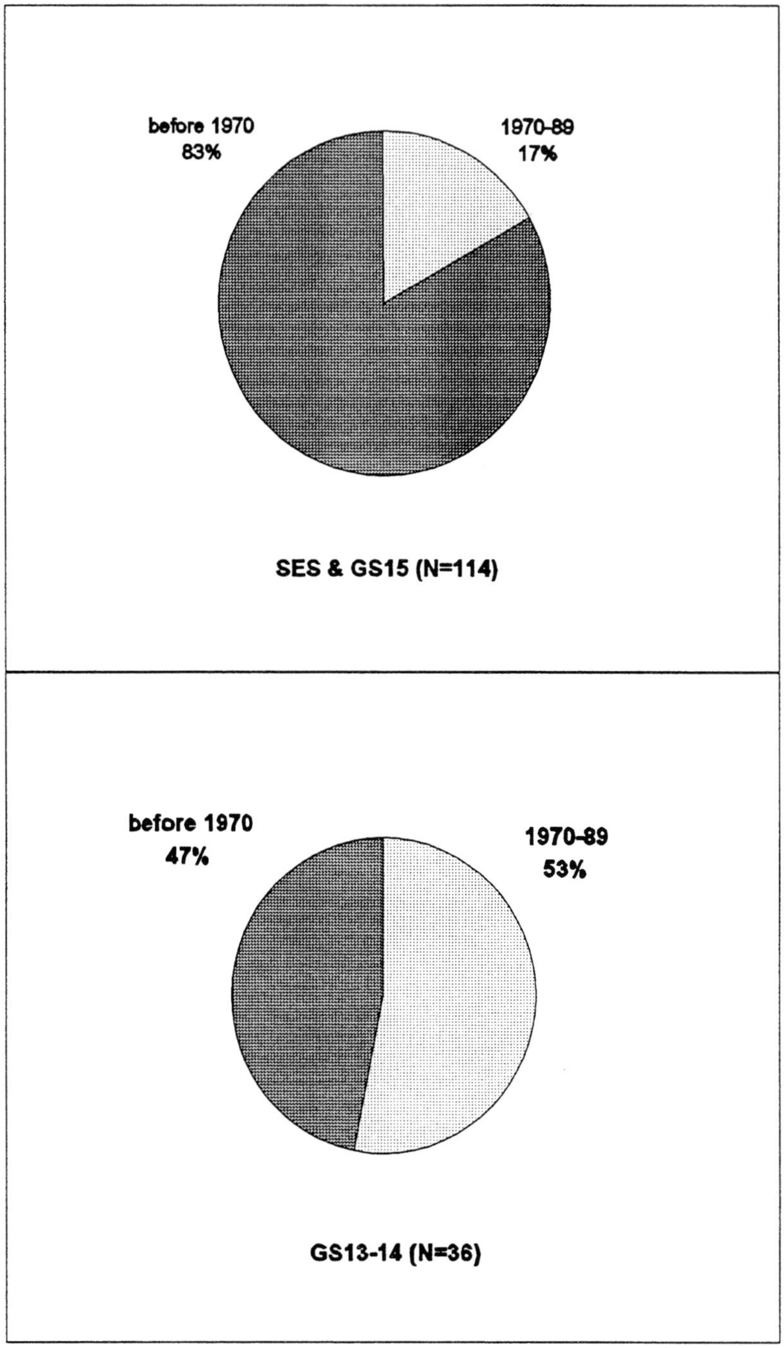

Source: Table A-5.2

FIGURE 2–12 ONR S&E work force year of college entrance.

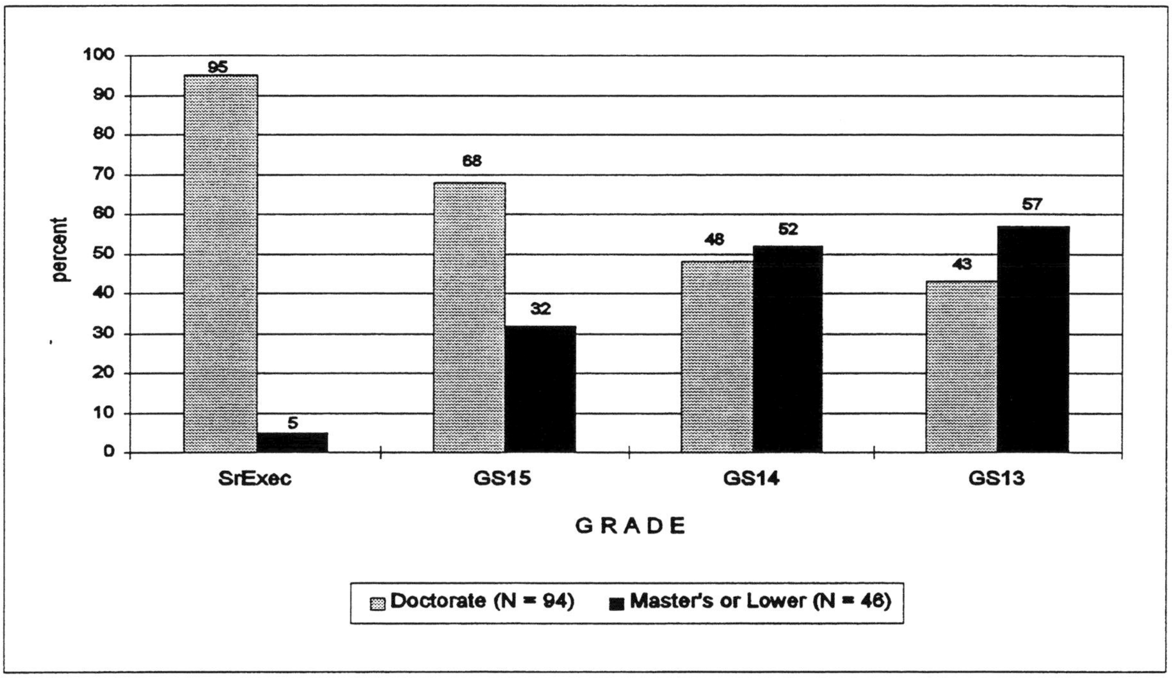

Source: Table A-5.3

FIGURE 2–13 ONR work force by grade and highest degree earned.

doctoral degree. Two-thirds of GS15s are educated to doctoral levels, though 11 (13 percent) have less than a master's degree. Almost half of the program officers at the GS13 and GS14 levels hold doctoral degrees. Overall, 63 percent of the male program officers and 53 percent of the female program officers in the data base have doctoral degrees.

Findings and Conclusions

Even with ONR's mixture of applied and basic research missions, the doctoral degree is still the preferred degree. With all but one senior executive and over half of the program officers, holding a doctoral degree, it is likely that promotional opportunities will be greatest among those with this advanced education. Women and minorities who are hired without this degree will have, in the committee's opinion, fewer opportunities for advancement (see section on "Recent Departures and Hires").

Employment Status

In order to compare ONR scientists and engineers with the national pool, the committee categorized them by five broad fields of employment, based on the occupational codes defined by the Office of Personnel Management (see note to Table A-5.1). The five fields are biological sciences, engineering, mathematics and computer sciences, and cognitive sciences. Data on the field of doctoral or master's degree for ONR employees were not available. It is important to note that the field of employment at ONR does not necessarily identify the division or department in which an individual works.

With 78 percent of the work force, the physical sciences and engineering clearly dominate the fields of employment, reflecting ONR's mission and its current portfolio of research and development activities (see Figure 2–14). Half of the senior executives work in the physical sciences (especially physics and oceanography); the other half are spread fairly evenly across the other four broad fields: mathematics and computer science, engineering, biological sciences, and cognitive sciences. Almost half (44 percent) of the program officers also work in the physical sciences, with another 36 percent in engineering. The remaining 25 people (20 percent) work in mathematics and computer science, biological sciences, and the cognitive sciences.

By contrast, the 16 women in ONR are spread across the five broad field areas as follows: 2 in the physical sciences, 5 in engineering, and 9 in the other three areas. Looking at each field area separately, 3 percent of the scientists and engineers in the physical sciences are women and 10 percent of those in engineering. By contrast, one-third of those in mathematics and computer science (5 out of 15) and in the biological sciences (3 out of 10) are women. Only one woman out of eight employees is identified as working in the cognitive sciences.

Table 2–1 illustrates the distribution of the S&E work force by grade. Three-quarters of the S&E employees work at the GS15 or senior executive level in each of the five field areas. Only half of the women occupy these ranks. By contrast, 5 percent of the overall work force are employed at the GS13 level, while 31 percent of the women (all hired between 1990 and 1995) are GS13s. As Figure 2–15 shows, women represent about 10 percent of all S&E employees at the GS14 and GS15 levels. They account for 63 percent, however, of the GS13 employees.

The overwhelming majority of scientists and engineers (140) started work at ONR since 1980. This includes 20 out of 23 of the senior executives, 94 percent of the male program officers, and all 16 of the women. Since 1990, seven of the senior executives have joined ONR, along with 36 (one-third) of the male program officers and 13 of the 16 women (see Table 2–2). Overall, 23 percent of the employees who

TABLE 2–1 Distribution of ONR Work Force Across Grades

Grade	Total		Male		Female	
Total	150	(100%)	134	(100%)	16	(100%)
GS13	8	(5%)	3	(2%)	5	(31%)
GS14	28	(19%)	25	(19%)	3	(19%)
GS15	91	(61%)	83	(62%)	8	(50%)
Sr. Executive	23	(15%)	23	(17%)	0	(0%)

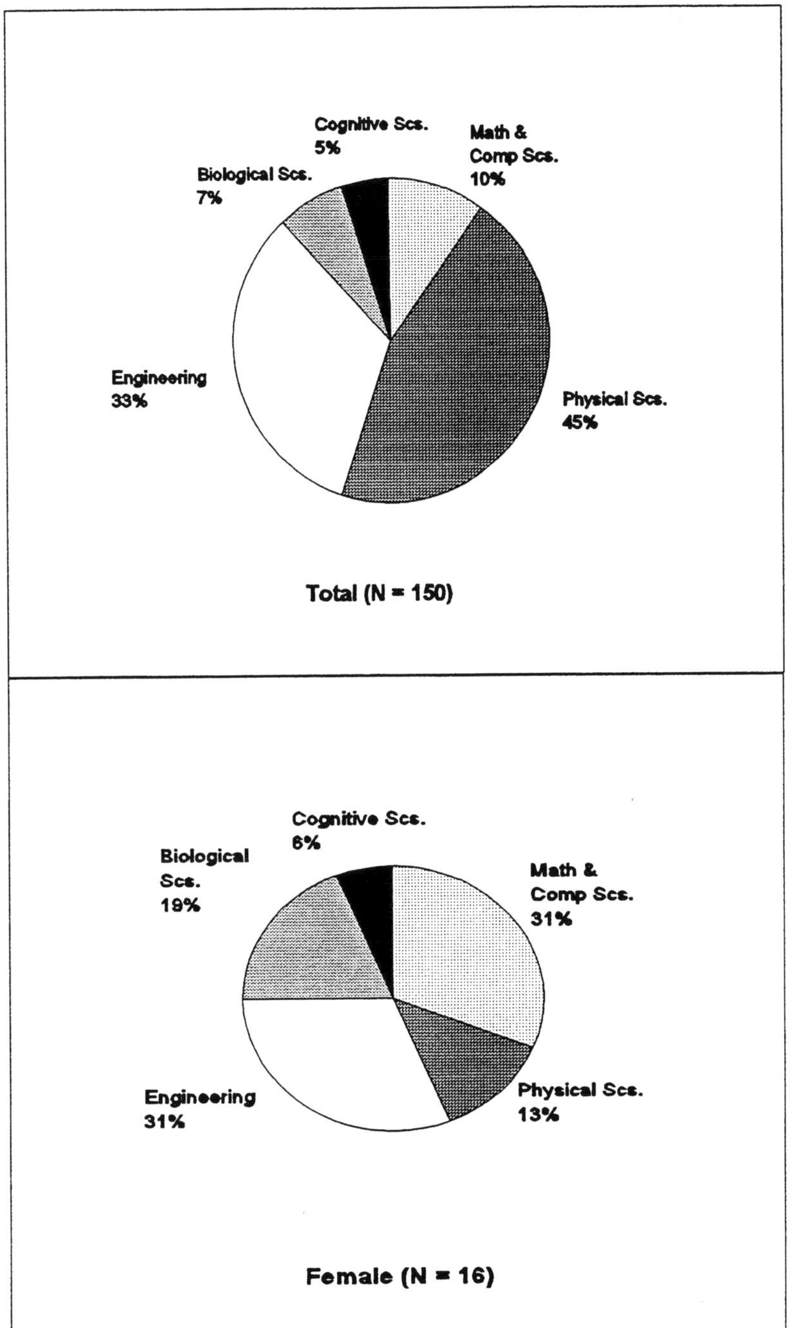

Source: Table A-5.4

FIGURE 2–14 Total and female ONR work force by field.

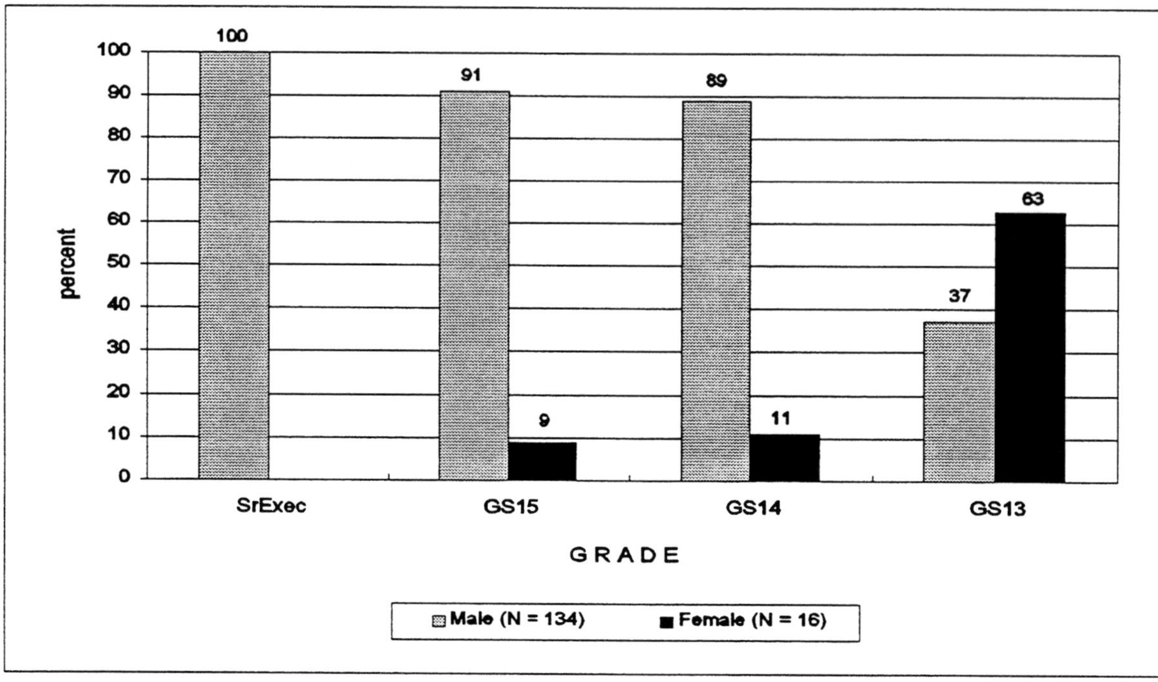

Source: Table A-5.4

FIGURE 2–15 ONR S&E work force by grade and gender.

began work at ONR since 1990 have been women, compared to 3 percent before 1990 (see Figure 2–16). ONR has thus made progress in hiring women in recent years, but few are at senior levels.

Promotional opportunities by gender are very difficult to assess since each case is different. Educational background, job performance, time in grade, and personal preference may all affect an individual's opportunity for promotion in ways which are difficult to measure. The differential data of "target grades" on male and female program officers at ONR are significant. An individual's target grade is the highest grade to which the person can be promoted from his or her current position without an open competition. The target grade for each

TABLE 2–2 ONR Hire Date by Gender

Grade	Total	1965–79	1980–84	1985–89	1990–95
Total	150	10	32	52	56
Male	134	10	30	51	43
Female	16	0	2	1	13

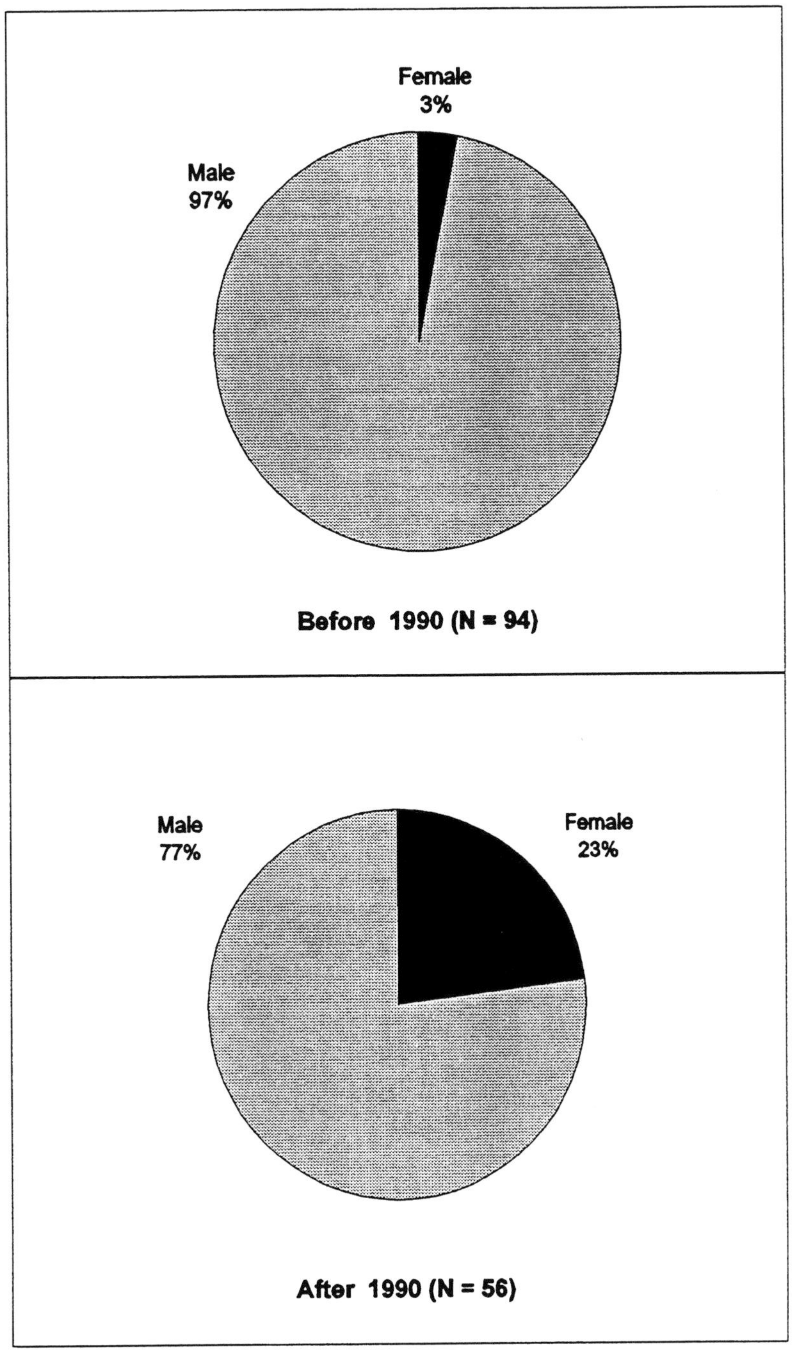

Source: Table A-5.5

FIGURE 2–16 ONR start date.

position is set at the time of recruitment, based on the job responsibilities, and is indicated by the highest grade listed in the job announcement. Individuals hired at less than GS15, for example, may or may not have GS15 as their target grade. For the GS13s and GS14s in this population, 16 out of 28 (59 percent) of the positions held by men are targeted at GS15, but only two out of eight (25 percent) of the positions held by women are targeted at GS15.

Women and men employed at the same grade level are similar to one another in terms of salary. The median salaries for both male and female program officers range from $60,000 at the GS13 level to $94,000 at the GS15 level. The median salary earned by female program officers overall ($83,500), of course, is lower than that earned by male program officers ($90,600), a reflection of the higher concentration of women at lower grades and possibly of shorter time in grade.

Findings and Conclusions

ONR has hired a number of female scientists and engineers since 1990; 23 percent of the new hires in the past six years have been women, including two in the physical sciences. The overall number of women, however, continues to be small, especially in the physical sciences where the largest number of employees work. Although the ONR program in the cognitive sciences is very small, the proportion of women is also low. This is surprising, given the facts that six out of eight ONR cognitive scientists have their degrees in psychology and that over one-third of experienced Ph.D. psychologists are women.

Employees are concentrated at the senior executive and GS15 levels. Women tend to be concentrated at the GS13 level, partly because of their shorter tenure at ONR, but also because of the number of them who have been hired in at that level. This fact, coupled with the smaller percentage of the women program officers who have target grades of GS15, suggests that promotional opportunities for women may be more limited.

Employment History

With help from ONR's Office of Human Resources, the committee was able to gather data on each individual's most recent and second most recent previous employers. Although, in the interviews, program officers stated that ONR scientists and engineers were recruited primarily from academia and industry, the data show that the majority came to ONR from elsewhere in the Navy. Seventy-nine percent of the current work force listed the federal government as their most recent employer prior to coming to ONR, and 61 percent listed the Navy (see Figure 2–17). Only 10 percent came directly from academia and 11 percent from industry. Data on each individual's second most recent employer show that nearly 70 percent came from the federal government, with 40 percent specifically from the Navy. Eighteen percent came from academia and 13 percent from industry.

The Navy was the most recent employer for 70 percent of the senior executives and 60 percent of the program officers. Of the female program officers, 13 out of 16 (over 80 percent) listed the federal government as their most recent employer, and half listed the Navy specifically. Ten specified the federal government as their second most recent employer.

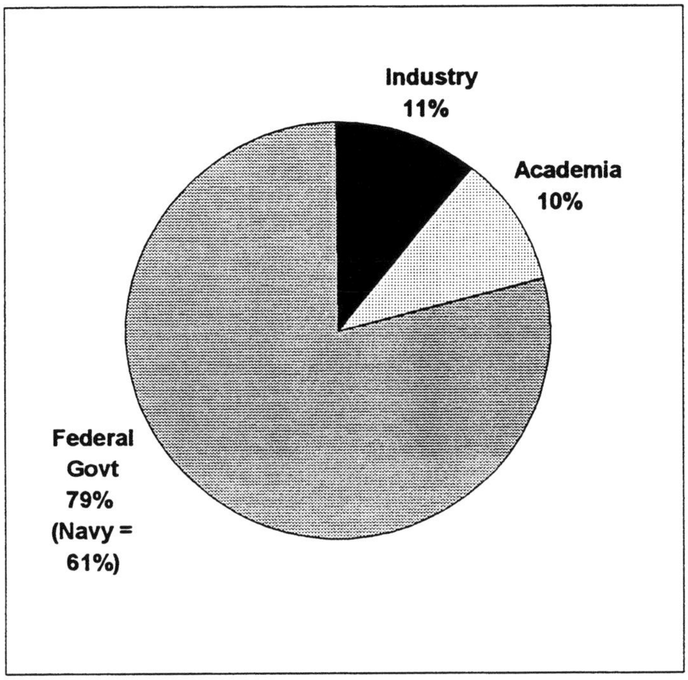

Source: Table A-5.8

FIGURE 2–17 ONR S&E work force by most recent previous employer.

Despite the predominance of Navy experience, there are a significant number of ONR scientists and engineers who have some work experience outside the Navy. Only 38 percent of the current ONR employees, for example, worked for the Navy in both of their previous employment situations. The remaining 62 percent provide experience and knowledge from a variety of employment sectors.

Findings and Conclusions

The committee noted that a high percentage of employees came to ONR from elsewhere in the federal government, and especially from the Navy. This is in contrast to the description of the ideal program officer offered by some ONR employees—a faculty member managing a research laboratory at a large university. The expectation of previous Navy experience is understandable; ONR is part of the Navy, and its work must contribute to solving the Navy's technical problems. At the same time, though, this expectation—whether explicit or implicit—limits ONR's recruitment possibilities, especially among underrepresented groups since three-quarters of those individuals work in academia or industry. The committee believes there might be ways to give new employees experience in and orientation to the Navy without expecting prior Navy or federal employment.

The majority of ONR employees who have some recent work experience outside of the Navy constitute a valuable resource for recruiting. Their knowledge of and contacts in academia and industry, or elsewhere in the federal government, could be very helpful in identifying qualified candidates for ONR positions from these other sectors.

Recent Departures and Hires

The committee was able to collect some information on employee departures. Based on records supplied by ONR, 87 scientists and engineers left ONR between 1988 and 1995. Seventy-eight of those individuals were white males, 5 were white females, 2 were African American males, and 2 were Asian American males. The number of departures has been fairly constant over this period—about 10 per year—with the exception of 1994 when that number doubled because of the reorganization of ONR. Assuming a population of at least 150 scientists and engineers during these years, the average annual attrition rate at ONR appears to be 6–7 percent.

The committee did not observe in these data a pattern of higher attrition among women or minorities than among white males, although the numbers are extremely small. Nevertheless, the committee suggests that tracking of attrition rates and exploring the reasons for employee departures through exit interviews can yield useful information about the work environment and any potential challenges in recruiting replacements.

From its reorganization in 1994 through March 1996, ONR hired 18 scientists and engineers: 11 men and 7 women. They were hired into physical sciences, mathematics and computer science, engineering, and biological science. For the physical sciences, the two women hires are the first to be employed as program officers.

The educational level of the men as a group is higher than that of the women, with eight males (73 percent) but only three females (43 percent) holding doctoral degrees. Four men were hired at the senior executive or GS15 level, compared with two women hired as GS15s. Only two men were hired at the GS13 level, while four women were, including two Ph.D.s (see Figure 2–18).

Like their colleagues, these new ONR employees came generally from prior Navy employment. All but one of the 18 recent hires listed the federal government as their most recent previous employer; one came from industry. Eleven out of the 18 (61 percent) came directly from the Navy. Thirteen (72 percent) reported the federal government as their second most recent employer, with two from academia and three from industry.

All of the female new hires came from federal employment. Five of the women had been already employed by the Navy and temporarily assigned to work at ONR at the time of their hire (often called "detailees"), including both women hired at the GS15 level. Three of the seven had worked in academia or industry previously.

Findings and Conclusions

Based on its review of the limited data available on recent employee departures, the committee concludes that turnover is not high among the science and engineering personnel at ONR. This impression is reinforced by comments made in the interviews to the effect that retention is not a large problem at ONR. There are a number of very long-term employees, as

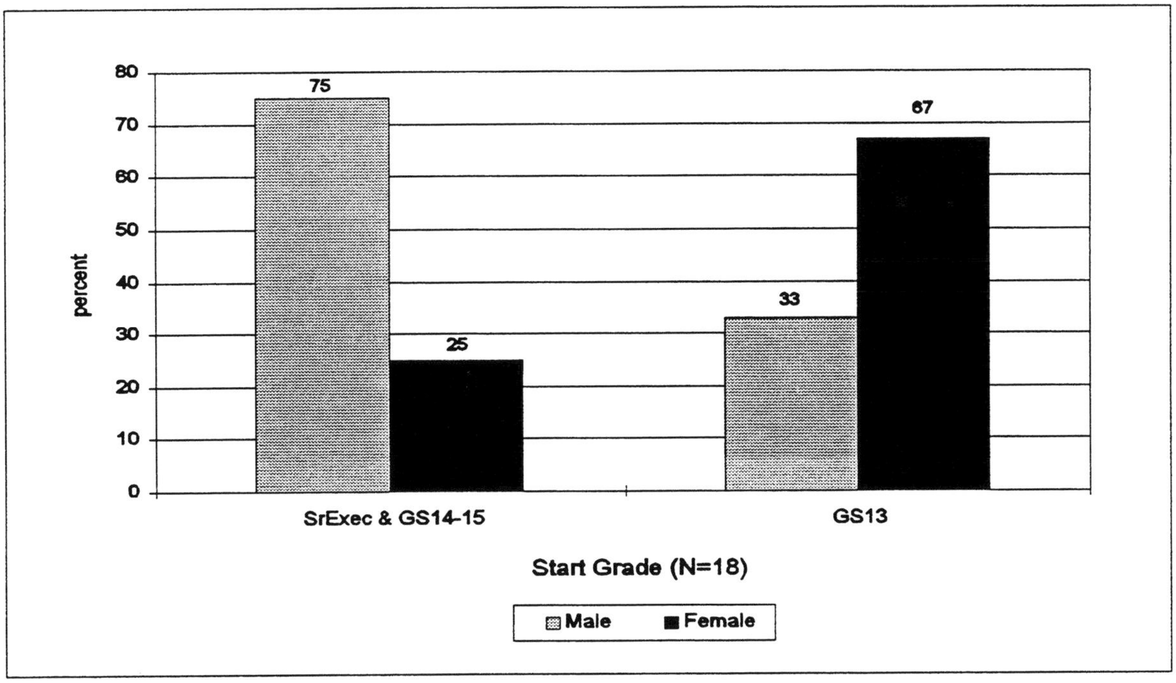

Source: Table A-5.10

FIGURE 2–18 ONR S&E recent hires by start grade and gender.

demonstrated by the data on start dates at ONR, and, once there, individuals tend to stay. The committee did not attempt to interview scientists and engineers who had separated from ONR in recent years, but such an activity, if done systematically, could yield important information about the work environment and suggestions for improvements.

With regard to new hires, ONR is to be commended for successfully recruiting seven women in the past two years, including two in the physical sciences, two in engineering, and three in mathematics and computer science. These new hires represent almost half of the entire female S&E work force. The committee notes that, as a group, these women were hired in at lower grades than the men, and more women were hired without the doctoral degree. Although the numbers may be too small to indicate a pattern, ONR needs to be careful not to concentrate new women into positions from which they are less likely to be promoted or rise to positions of leadership.

The picture for minorities is not so positive. There is only one African American and no Hispanic or American Indian in the work force, and none of the 18 recent hires was a member of a minority group. Although the number of minorities with appropriate backgrounds for these positions is small, they do exist, and the committee could not find any evidence that ONR has yet risen to the challenge of finding and making the extra effort to recruit them.

OTHER POOLS

Naval Research Laboratory Personnel

Because so many ONR employees come from the Navy, the committee believes that a potential pool of experienced scientists and engineers who also have Navy experience exists in the Navy laboratories, and especially at the Naval Research Laboratory (NRL). Time did not permit an extensive investigation into the various Navy research facilities, but the committee did review some basic demographic and employment data on women and minorities at NRL. Individuals at the GS12 level were included in this analysis since they could potentially be eligible for positions at ONR as program officers.

Out of 1,677 scientists and engineers at the GS12 through senior executive levels, 170 or 10 percent are women (see Figure 2–19). The physical sciences and engineering clearly dominate NRL's activities with 90 percent of the S&E work force. Three-quarters of the female scientists and engineers work in the physical sciences or engineering.

There are 44 individuals (less than 3 percent of the total) who are members of underrepresented racial or ethnic groups: 22 African Americans, 16 Hispanics, and 6 American Indians. Thirty-eight (or 86 percent) of these scientists and engineers work in the physical sciences and engineering, with the remaining 6 in mathematics and computer science.

NRL appears to have a relatively large population of disabled scientists and engineers (nearly 8 percent or 130). However, this includes everyone who reports a disability, which is a more inclusive definition than that used in the national surveys of doctoral and master's recipients.* Without further investigation, it is difficult to compare these data on disabled persons with those from other sources.

In terms of grade level, female scientists and engineers are generally represented in fewer numbers as their grades increase, ranging from 17 percent (80) of the GS12s to 2 percent of the GS15s (6) (see Figure 2–20). The senior executive ranks, however, are an important exception to this pattern; there are 4 women at this level, all in the physical sciences. These senior women, combined with their 6 female colleagues at the GS15 level, would appear to be a potential source of recruitment for ONR senior executive positions.

Underrepresented minorities also populate the lower grades in greater numbers, with 31 (or 70 percent) of them in GS12 or GS13 levels. There are, however, 5 individuals at the GS15 or senior executive level who might be eligible for senior positions at ONR.

---

*Both the Survey of Doctorate Recipients and the National Survey of College Graduates define disabled persons as "individuals who have severe difficulty seeing, hearing, walking, and/or lifting or are unable to perform these tasks." The Survey of Earned Doctorates defines disabled persons as "those individuals who indicated that they had a disability."

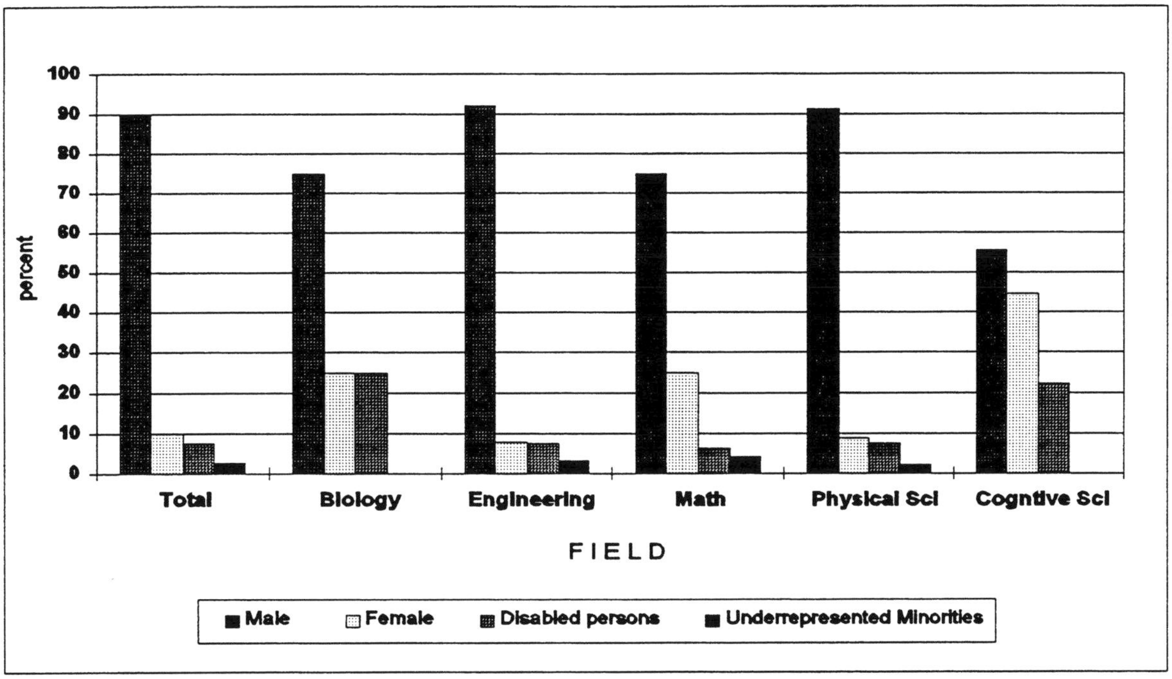

Source: Table A-6.1

FIGURE 2–19 NRL S&E employees by field, gender, race, and disability status.

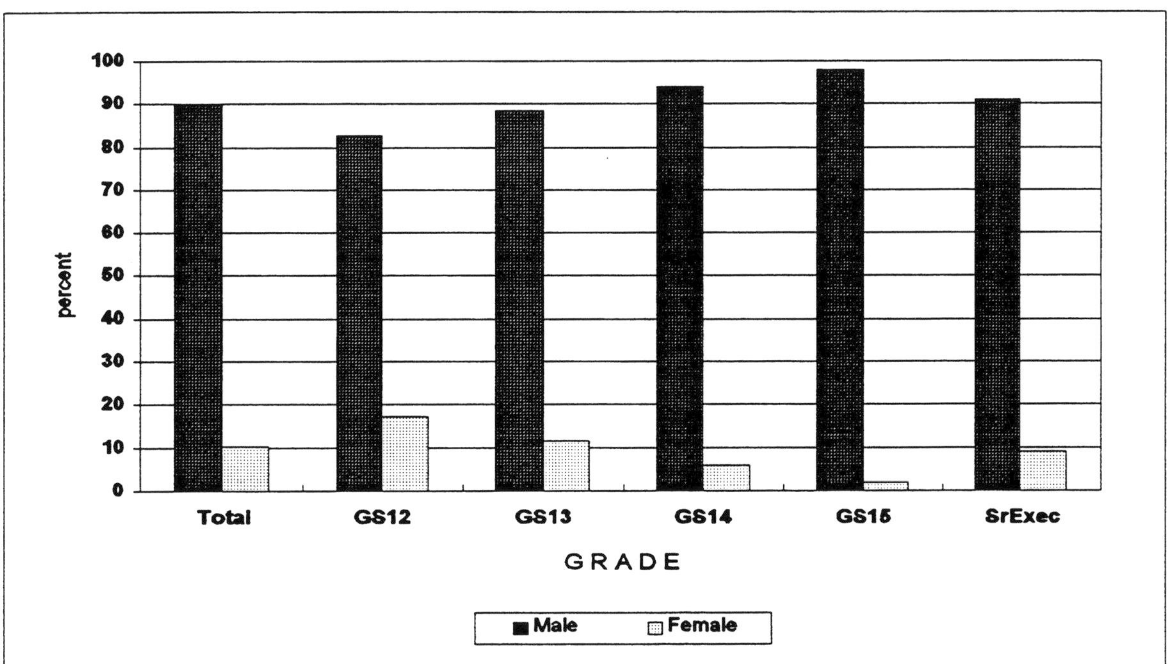

Source: Table A-6.2

FIGURE 2–20 NRL S&E employees by grade and gender.

ONR Principal Investigators

There are approximately 5,400 principal investigators (PIs) with contracts or grants funded by ONR. These individuals have an ongoing, working relationship with ONR program officers and, by definition, are engaged in work of interest to the Navy. They constitute, in the committee's opinion, a potential source of employees for ONR that does not appear to have been tapped. They might come to ONR as permanent employees or through the Intergovernmental Personnel Act (IPA), by which employees of organizations outside the federal government undertake short-term assignments in federal agencies.

Unfortunately, the data on principal investigators do not contain information on their gender, race/ethnicity, or disability status. A simple review of the names of the PIs, however, yielded a rough estimate that approximately 525 (or 10 percent) of them are women. Over half of these women were located in academia, with at least another third in government (see Figure 2–21).

Findings and Conclusions

Both NRL employees and ONR principal investigators can serve as potential pools for the recruitment of more women and minorities into ONR program officer

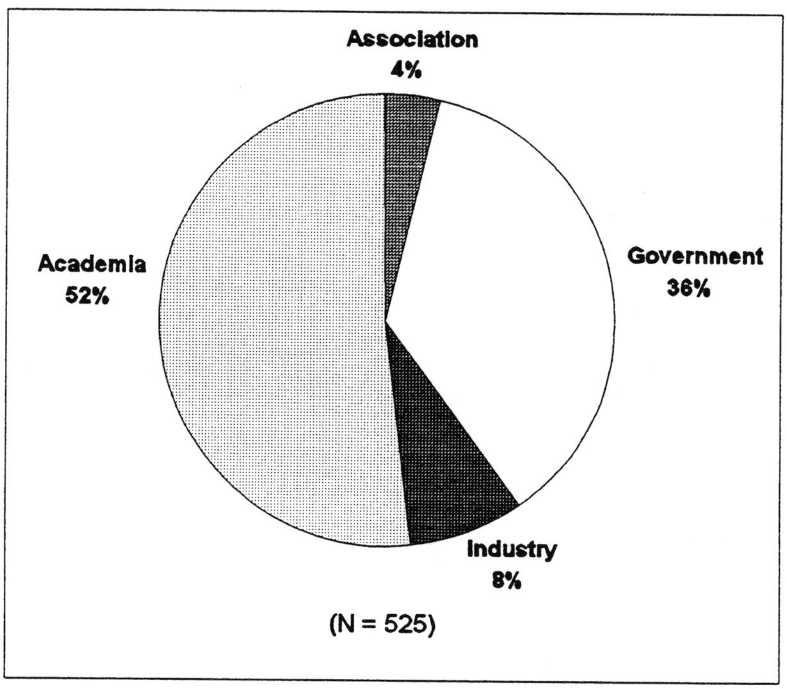

Source: Human Resources Office, Office of Naval Research, 1996

FIGURE 2–21 Female principal investigators by sector of employment.

and management positions. Most have the necessary technical background, as well as desirable experience in or with the Navy. Many may not be interested in giving up their research careers, of course, but no doubt some would be. The population at NRL seems particularly promising, especially if recruitment efforts are focused on those women, minorities, and disabled persons with doctoral degrees.

COMPARISON OF ONR EMPLOYEES WITH THE NATIONAL POOLS

There are a significant number of women and minorities, and a growing number of persons with disabilities, among scientists and engineers who are or soon will be eligible for positions at ONR. Given the age, fields, and background of the majority of ONR employees, however, it is not surprising that few of them are members of these underrepresented groups.

First, the percentage of women, minorities, and persons with disabilities in the age cohort of most ONR employees is small. Figure 2–22 compares the year of college entrance of ONR employees with doctoral degrees to that of the national pool of experienced Ph.D.s (comparable data were not available for master's recipients). Among Ph.D. scientists and engineers who entered college before 1970, the proportion of women in the national pool does not exceed 15 percent. For each age cohort, the proportion of women in the ONR work force is still less than that in the national pool, but the ONR percentages are comparable to those in the national pool for 1960–69 and 1970–79.

Second, proportional representation by these groups has historically been small in the fields of greatest interest to ONR, namely the physical sciences and engineering. Among experienced Ph.D.s, the broad fields with the lowest percentages of women are the physical sciences (9 percent) and engineering (3 percent). Among experienced master's in engineering, women account for 6 percent.

Third, many ONR employees come from within the Navy or elsewhere in the federal government, which is true for only a small percentage of the target group. As the data on the available pools indicate, three-quarters of the underrepresented Ph.D.s who are likely to be eligible for ONR positions currently work in academia or industry, as do about 76 percent of the master's-educated engineers. Less than 8 percent of these Ph.D.s, and 12 percent of the master's in engineering, work in the federal government.

Given these limitations, the committee concludes that ONR has done a reasonable job of hiring women into its science and engineering work force. It has been less successful in hiring minorities, but its number of persons with disabilities exceeds the percentage in the overall pools. Each of these groups will be examined in turn.

Representation of Women

Table 2–3 compares the number and percent of women currently in the ONR S&E work force with the number and percent in the various national pools by field. In examining these data, it is important to remember that, given the small numbers involved, percentages (especially in the ONR work force of 150 people) must be used with caution. It would be unreasonable to expect ONR to attempt to match the percentage of women (or minorities) in each

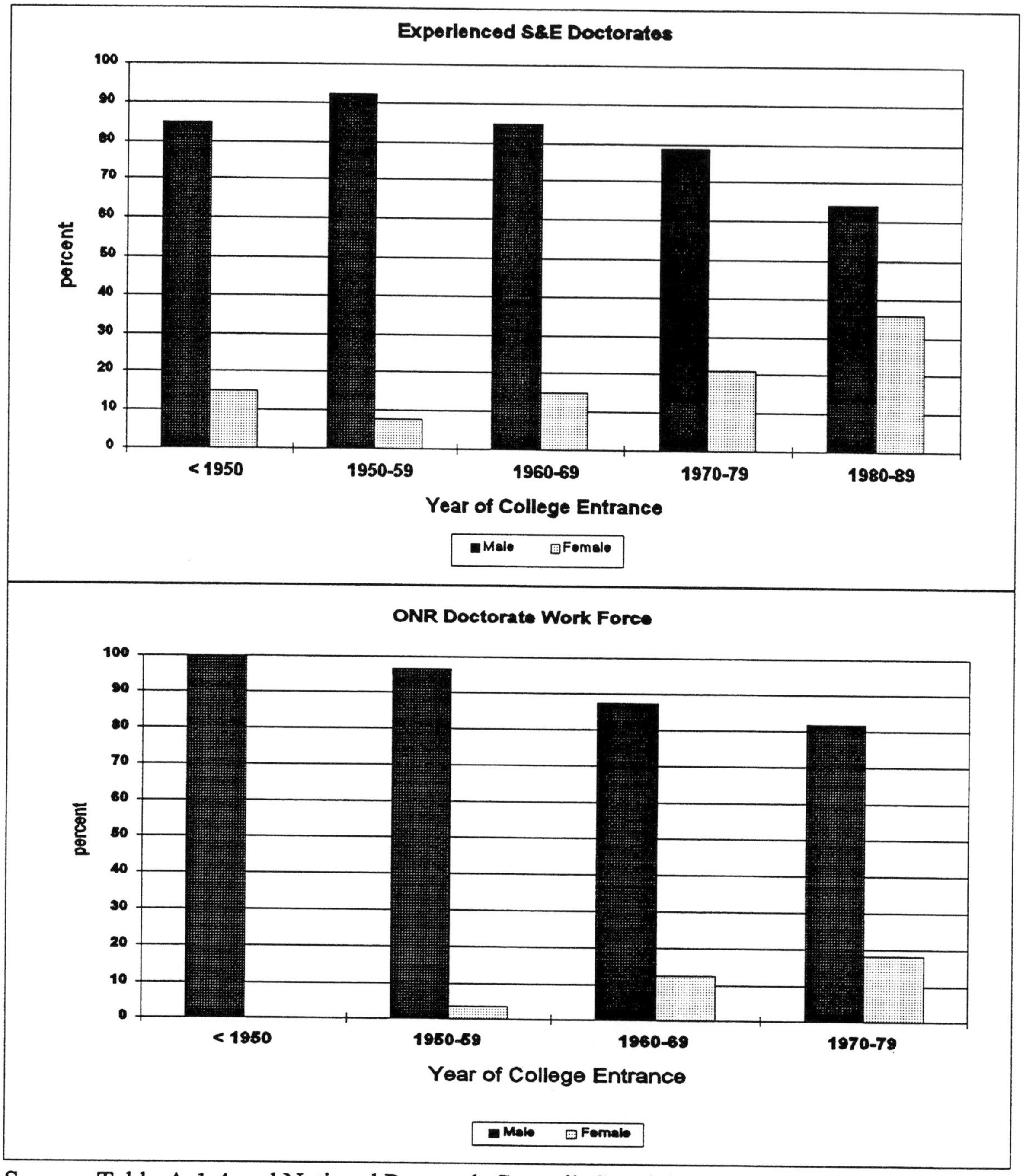

Source: Table A-1.4 and National Research Council, Special Tabulations, 1996.

FIGURE 2–22 National pool of experienced S&E and ONR doctorate work force by year of college entrance and gender.

TABLE 2–3 Percent of Women in the ONR Work Force and National Pools by Broad Fields

Group	Total		Biological Sciences		Engineering		Mathematics & Computer Sciences		Physical Sciences		Cognitive Sci.* & Psychology	
	#	%	#	%	#	%	#	%	#	%	#	%
ONR	16	10.7	3	30.0	5	10.0	5	33.3	2	3.0	1	12.5
Experienced												
Doctorate	33,553	14.7	16,246	23.7	1,653	3.3	2,159	11.2	5,925	8.6	7,570	34.0
Master's	17,865	6.2	--	--	17,865	6.2	--	--	--	--	--	--
Recent												
Doctorate	16,660	30.2	7,759	41.1	1,618	14.6	1,063	22.2	3,216	21.7	3,004	55.2
Master's	7,370	13.2	--	--	7,370	13.2	--	--	--	--	--	--

NOTE: The national pool for cognitive sciences includes Ph.D.s in the following fields of psychology: cognitive and psycholinguistics, comparative, experimental, industrial and organizational, personality, physiological, psychometrics, quantitative, social, and general.

SOURCE: Tables A-1.1, A-2.1, A-3.1, A-4.1, A-5.1

of its fields to the percentage in the national pool. In addition, a suitable applicant for employment at ONR in a particular group of fields (e.g., the physical sciences) will not necessarily have a degree in one of those fields. Scientists in many areas, and engineers especially, tend to work in fields different from that of their graduate education. With these caveats, the committee believes that Tables 2–3, 2–4, and 2–5 can help identify potential areas of underutilization.

How well is ONR doing at achieving representation of women, by field, compared with national pools? The box below summarizes the committee's findings.

The agency-wide representation of female scientists and engineers at ONR is good considering the age and experience of the current work force. ONR has clearly made efforts in the past few years to recruit women, and the ONR work force is now nearly 11 percent female, comparable to the percent of women in the national pools of experienced scientists and engineers (6–15 percent). However, more effort is needed to recruit or promote women into the higher management levels. Also, effort should be focused on the physical sciences where the representation of women at ONR still lags behind the pool of experienced female Ph.D.s. The gap appears to be easier to close in the cognitive sciences, given the larger representation of women in that pool.

Additional effort in the physical sciences and the cognitive sciences should pay off in the next five years, especially with the percentage of women among recent

FIELD	FINDING	RATIONALE BEHIND FINDING
Biological Sciences	meets a reasonable goal	Thirty percent of ONR biologists are women. This is better than the national pool of experienced Ph.D. biologists (24 percent), though below the more recent supply (41 percent).
Engineering	meets a reasonable goal	Ten percent of ONR engineers are women, far better than the 3–6 percent in the experienced pools of Ph.D.s and master's recipients and almost at the 13–15 percent level of the recent graduates.
Math and Computer Science	exceeds a reasonable goal	Thirty-three percent of ONR mathematicians and computer scientists are women. This far exceeds the 11 percent of women in the experienced Ph.D. pool and even the 22 percent in the recent graduate pool.
Physical Sciences	needs improvement	Three percent of ONR physical scientists are women, while women constitute 9 percent of the experienced Ph.D. pool and 22 percent of the recent graduates.
Cognitive Sciences	needs improvement	Twelve percent of ONR cognitive scientists are women, compared to 34 percent in the experienced pool and 55 percent of recent graduates.

graduates nationally ranging from 13 to 30 percent. For example, assuming ONR maintains its current size, the hiring of an additional 10 women would bring its overall percentage of female scientists and engineers to 17 percent, comparable to the 15 percent in the pool of experienced Ph.D.s and the 30 percent of recent doctorates.

Separate efforts will need to be made to recruit or promote at least two or three women (9–13 percent) into the senior executive ranks, regardless of field, and to help ensure that women employees in all fields have opportunities for promotion comparable to those of their male colleagues.

Representation of Minorities

Because of the presence of only one underrepresented minority at ONR, little can be said about the comparison of the ONR work force with the national pools. Table 2–4 does provide the percent of underrepresented minorities in each of the national pools by field, which can be used as a guide to future recruitment. Across all fields, a minority population at ONR of 5 to 6 scientists and engineers out of 150 would approach the 3–4 percent in the overall pools.

Representation of Persons with Disabilities

Table 2–5 describes the number and percent of persons with disabilities in each of the national pools. As with minorities, the number of persons with disabilities at ONR (three) is too small to permit comparisons by field. The overall percentage of persons with disabilities in the ONR work force is 2 percent, which is higher than the 0.5 percent in the pool of experienced doctorates and the 1.2 percent among recent doctorates.

Summary

The human resource pool from which ONR can draw is larger and more diverse than in the past. There are today a significant number of women and minorities who are experienced scientists and engineers. Also, their representation in the cohort of recent doctoral and master's recipients is growing. There are women and minorities at NRL who would be eligible for positions at ONR and who have the advantage of Navy experience, as well as female researchers who are already ONR principal investigators. These pools are rich resources from which to draw as ONR continues to increase its diversity.

TABLE 2–4 Percent of Underrepresented Minorities in the ONR Work Force and National Pools by Broad Fields

Group	Total		Biological Sciences		Engineering		Mathematics & Computer Sciences		Physical Sciences		Cognitive Sci.* & Psychology	
	#	%	#	%	#	%	#	%	#	%	#	%
ONR	1	0.7	0	0.0	1	2.0	0	0.0	0	0.0	0	0.0
Experienced												
Doctorate	7,343	3.2	2,248	3.3	1,403	2.9	661	3.4	1,857	2.6	1,173	5.3
Master's	5,666	2.0	--	--	5,666	2.0	--	--	--	--	--	--
Recent												
Doctorate	2,416	4.5	834	4.4	459	4.1	135	2.9	555	3.8	433	7.9
Master's	2,386	4.3	--	--	2,386	4.3	--	--	--	--	--	--

NOTE: The national pool for cognitive sciences includes Ph.D.s in the following fields of psychology: cognitive and psycholinguistics, comparative, experimental, industrial and organizational, personality, physiological, psychometrics, quantitative, social, and general.

SOURCE: Tables A-1.1, A-2.1, A-3.1, A-4.1, A-5.1

TABLE 2–5 Percent of Disabled Persons in the ONR Work Force and National Pools by Broad Fields

Group	Total		Biological Sciences		Engineering		Mathematics & Computer Sciences		Physical Sciences		Cognitive Sci.* & Psychology	
	#	%	#	%	#	%	#	%	#	%	#	%
ONR	3	2.0	--	--	--	--	--	--	--	--	--	--
Experienced												
Doctorate	1,484	0.6	289	0.4	264	0.5	199	1.0	517	0.8	216	1.0
Master's	1,292	0.4	--	--	1,292	0.4	--	--	--	--	--	--
Recent												
Doctorate	664	1.2	218	1.2	119	1.1	64	1.3	186	1.3	77	1.4
Master's	0	0.0	--	--	0	0.0	--	--	--	--	--	--

NOTE: The national pool for cognitive sciences includes Ph.D.s in the following fields of psychology: cognitive and psycholinguistics, comparative, experimental, industrial and organizational, personality, physiological, psychometrics, quantitative, social, and general.

SOURCE: Tables A-1.1, A-2.1, A-3.1, A-4.1, A-5.1

3

BUILDING A DIVERSE WORK FORCE

To build a more diverse work force, an organization needs to enhance two critical aspects: the process by which employees are recruited and hired, and the environment in which they work. Creating and maintaining a supportive and productive work environment and providing opportunities for employee development are as important in retaining quality employees as is the process of recruiting them.

RECRUITMENT AND HIRING

The committee gathered information on ONR's recruitment and hiring practices from several sources. The Office of Human Resources within ONR provided extensive background information and sample materials describing current recruitment and hiring policies and practices. They also permitted examination of 13 files documenting recent hiring actions of scientists and engineers into ONR. In addition, personal interviews with program officers and senior executives yielded important insights into the attitudes and beliefs that underlie hiring decisions.

The Current Process

A division or department head, otherwise known as the selecting official, initiates the hiring process and obtains approval from management and the ONR Office of Human Resources. Beyond the required vacancy announcements, the program officer or division head is responsible for deciding where to advertise. Once applications are received, the selection involves decisions at five levels:

Preliminary Review

A staffing specialist and a subject matter expert examine each applicant's file to determine who is qualified. For each qualified applicant, the staffing specialist then determines the grade level based on the applicant's prior experience and other qualifications.

Rating and Review Panel

The list of qualified applicants and their supporting documents are forwarded to a rating and review panel, usually appointed by the selecting official. Panel members rate each applicant separately and then meet as a group to try to reach consensus on any numbers on the rating sheet that have a significant variance. The panel determines the cut-off points for "best qualified" and "qualified" applicants and decides whom to refer to the selecting official.

Selecting Official

The selecting official decides whether to interview and, if so, whom to interview. The official can conduct the interview(s) alone or invite others. After the interview(s) the selecting official decides whom to recommend for hiring and prepares a letter of nomination.

Science and Technology Advisory Board (STAB)

The proposed candidate makes a presentation to the STAB, which is composed of the department heads (or their designated representatives) plus the Deputy Chief of Naval Research. STAB makes a recommendation to the Deputy Chief of Naval Research.

Final Approval

The Deputy Chief of Naval Research makes the final hiring decision.

Recent Hiring Actions

An analysis of recent hiring actions in ONR provides some useful insights into how the process works. As indicated above, the committee reviewed detailed data compiled from 13 "case files" of hiring actions for scientists and engineers completed between January 1, 1993, and December 31, 1995. (Summary data on these 13 case files are found in Appendix B.)

Among these 13 cases the process of advertising and recruitment was highly variable. In two cases, there was no recruitment: no vacancy announcement was prepared or advertised, and the individual selected (in both cases a male) was the only candidate. In four cases, applications were limited to current Navy or DoD employees, and three out of four of those were advertised only in the Washington, D.C. metropolitan area. The remaining seven cases showed evidence of open recruitment (outside the DoD), and six of the openings in this group were advertised nationally. Of the 13 cases, two showed evidence of having been advertised in journals or with associations of particular interest to women and minorities.

Of the nine cases where the job vacancy was advertised and the number of applications is known, the committee was able to identify a total of 210 applications, or about 23 per position. The actual number of applicants ranged from less than five for positions open only to Navy personnel in the local region to 50 for nationwide searches. Of the 210 applicants, 34 or 16 percent were female. No data were available on the race or disability status of the applicants since collection of that information is not required.

As indicated above, the hiring process at ONR requires that a rating and review panel examine the applications of all "qualified" candidates, rate them, and make recommendations to the Selecting Official. Because this is such a critical step in the selection process, the committee was especially interested in the composition of these panels and the actions of the selecting officials. In nine of the hiring actions, a rating and review panel of two to five people had been used. Three of the nine panels included a female panel member, one of whom was from outside ONR headquarters. The data do not reveal whether any panel member was disabled or came from a minority group. To the committee's knowledge, all Selecting Officials were white males.

Opinions About Recruitment and Hiring

In their interviews, senior executives and male program officers expressed similar views about the recruitment and hiring of minorities and women (see Appendix C). Many stated that ONR was "bending over backwards" to promote diversity, but that there was an inadequate supply of minority and female applicants nationally. Others expressed concern that hiring more minorities and women would dilute the strength of the ONR work force and lower quality. Some senior executives explained that women and minority senior scientists and engineers who are "stars" in their fields will not want to give up work in their field to be administrators or to work for the federal government, and that most of those qualified to work at ONR are competed for successfully by industry or academia.

Some senior executives believed that diversity will strengthen ONR and that more aggressive measures need to be taken to recruit and hire women and minorities. These individuals cited examples of the beneficial effects of diversity within their own units.

Female program officers were generally more skeptical about ONR's recruitment efforts. While they agreed that there were few female applicants for positions, some believed that an adequate effort to recruit outside the organization was rarely made. Others observed that women do not apply for senior-level jobs at ONR because they believe the positions are "hard-wired" for men and that women's careers do not always fit the traditional male molds. None of the women interviewed reported having served on an ONR rating and review panel, although several of their male counterparts did.

Findings and Conclusions

Although the number of recent hiring actions that the committee was able to examine is small, they do suggest several areas of concern. Recruitment efforts appear to be very uneven and quite dependent on the preference of the selecting official, including whether there will be recruitment at all, how open it will be, and how much advertising will be done. In the two cases examined where there was no recruitment and a white male was hired, there was no opportunity for women or minorities even to apply. Positions that were advertised only in the Washington, D.C. area or restricted to individuals already in the Navy had little chance of attracting a diverse pool of applicants, especially for senior positions.

Similarly, the committee could find little evidence of concerted efforts to bring women or minorities into the recruitment pools. Advertising was generally limited to a few standard venues, some of which are mandated.

Given the current job market, the total number of applications for many of these positions was surprisingly low. In the experience of committee members, the hiring of scientists and engineers in industry or academia may generate several hundred applications. Recent hiring actions in the Characterization Science and Services section of Corning, Inc., for example, yielded 600 applicants for two Ph.D. chemist positions. In 1995, the University of Virginia received 142 applications for a faculty position in structural engineering, and the University of California at Berkeley received 179 applications for an assistant professorship in chemistry. The University of Illinois at Urbana-Champaign reports receiving 250–400 applications per year for typically a single opening in computer science. The number of applications for academic positions, of course, may be inordinately high, given the large number of Ph.D.s seeking such employment.

Although three of the nine rating and review panels contained a woman, none contained a minority or a person with disability, to the committee's knowledge, and none of the ONR female program officers said they had served on any of these panels. The committee was also struck by the authority vested in the selecting official who can decide whether or not to interview anyone, whom to interview, and whom to recommend. Although a single manager may often make the hiring recommendations, the process by which those decisions are made should be controlled by appropriate checks and balances. The committee also believes that the STAB should make its hiring recommendations in the absence of the Deputy Chief of Naval Research since that individual makes the final hiring decision.

The picture drawn by these hiring actions is not consistent with the view expressed by some senior executives and program officers that ONR is "bending over backwards" to promote diversity. The concern that minority and women "stars" especially would not come to work at ONR is impossible to assess since the committee found no evidence that any efforts had been made to recruit individuals in this category. If senior ONR managers believe that ONR is not now competitive with academia or industry for top scientists and engineers, they may want to develop ways to change that impression. The belief held by some that increased diversity means lower quality will inhibit their ability to recruit effectively.

ONR has made progress in recruiting a more diverse work force, but the committee believes there are many resources that have not yet been tapped. Although substantial effort is required in the recruitment of minorities and women, such an effort can be successful.

THE WORK ENVIRONMENT

The committee's understanding of the work environment for ONR scientists and engineers is based primarily on answers to the survey questionnaire and the individual interviews summarized in Appendix C. Comments on the most significant aspects of the findings follow.

Almost everyone reported working long hours, with many working over 60 hours a week; the senior executives reported the greatest number of hours worked. Nearly everyone at every level experienced stress; they felt they are overworked and lack adequate staff support. Nevertheless, most reported that ONR is the best job they have ever had.

In general, the women were less satisfied than the men, except in the area of income. Many women believed they are treated as second-class citizens, and some saw the work environment as hostile. Some believed they control fewer research dollars, even within the same disciplinary groups.

Long-term female employees were generally less pleased than recently hired female employees. Some of the women with Ph.D.s were the least pleased, as they saw few professional growth opportunities, less use of their abilities, and less harmony between their personal and professional lives.

Many program officers reported not understanding the criteria for promotion. Some women were interested in promotion to the senior executive level but were not hopeful. They suggested that being an "acting head" of a division might be a route to promotion but that few women had held those positions.

Male program officers, while sharing some of the complaints of their female colleagues, generally expressed a sanguine attitude toward the organization. Like senior executives, many believed that ONR was doing what it could to increase diversity, and that there was no overt discrimination. The male program officers differed from the senior executives, however, in expressing during the interviews more negative stereotypes about non-whites and women.

In general, white males saw the environment as supportive. Senior executives were more satisfied with their jobs than were the program officers. They were aware the environment is competitive, but they were unaware that women see the work environment as hostile. Most saw no difference in the treatment given to women and minorities from that given to them.

A number of program officers, especially women, described an atmosphere in job interviews, briefings, meetings, and competition for funds that was "aggressive," "hostile," "combative," and "divisive." They stated that this atmosphere was unsupportive and unproductive, and some said that they were denied opportunities to give briefings presumably because their style was not aggressive enough.

While senior executives agreed that diversity would strengthen the scientific enterprise by introducing divergent perspectives and experiences, many were emphatically committed to the adversarial or confrontational approach. They believed that this approach is necessary to argue successfully for one's budget priorities in an era of shrinking resources and in a critical and sometimes hostile bureaucratic environment. ONR leadership seemed especially committed to the adversarial interview for job candidates, asserting that this kind of hurdle is essential to finding the right people. Some senior executives noted that women do at least as well as their male counterparts in challenging and confronting others, and they insisted that the climate is the same for males and females, minorities, and non-minorities—"hard on everyone." Other senior executives did not acknowledge that such an atmosphere even exists.

The committee identified some other aspects of the ONR work environment which, though not directly related to the diversity issues under discussion here, do affect the work environment and productivity of ONR employees. Some non-white males cited examples of where they felt they were not given the same opportunities as white males. Also, a number of employees from the former Office of Naval Technology (ONT) and Office of Applied Technology (OAT) believed they were treated as second-class citizens, reinforcing the tension between the culture of basic research (6.1) and applied research and development (6.2 and 6.3).

The current ONR culture is dominated by basic research and by the Ph.D. degree; 22 out of 23 senior executives hold the doctorate, and 17 were managers of predominantly basic research programs prior to the merger of ONR, ONT, and OAT. At the same time, the Navy is promoting more application of research to its technical needs, and applied research and development funds represent over 70 percent of ONR's annual budget. In the interviews, program officers, but especially senior executives, identified differences in education, knowledge of the Navy, and research philosophy as divisive issues to be overcome in the new, consolidated ONR.

Attitudes Toward Diversity

There are divergent views on diversity within ONR. Most employees recognized that with the downsizing of ONR, diversification will be difficult because there are few opportunities to hire. Among senior executives, diversity was an issue that had to do almost exclusively with recruitment and hiring, rather than with the work environment.

Many male program officers and senior executives alike believed that ONR is doing what it can to diversify its work force, as discussed in the section on "Recruitment and Hiring." By contrast, female program officers pointed out numerous aspects of ONR's current work environment and climate that were not conducive to the retention and development of minorities and women.

Findings and Conclusions

The work environment for scientists and engineers at ONR is dominated by individuals who are white, male, and have a long association with the Navy. It is also dominated by scientists with doctoral degrees. Given the history of ONR and the long tenure of some of its senior personnel, this environment is what one might expect, and this kind of work force has served ONR well for many years. It may not be serving ONR's current needs as well, however. A number of scientists and engineers do not relate to the prevailing management style and perceive that they have a difficult time at ONR. They clearly like their work, but they do not see the opportunities for advancement and support that allow them to use their full capacities. This may have a negative impact on ONR's ability to recruit high quality people.

Some of the differences in perceptions between men and women are striking. While these perceptions can in no way be attributed to every woman or every man at ONR, they were expressed by enough members of each group to be noteworthy. Whether fact or perception, such significant differences in employees' views of the same organization are counterproductive and undermine the ability of the agency to function as an integrated unit.

Providing a work environment that is supportive of all employees, not just those in the dominant groups, is critical to productivity. For S&E organizations that have been successful in recruiting and retaining professional women, a key element appears to be some type of support system. "Buddy systems," where an individual is matched formally or informally with a more senior employee, are used by Corning,

Xerox, Apple, Aerospace, and several laboratories of the Department of Energy to create a supportive environment for women. The buddy system can provide mentoring on a one-to-one basis or make available a "council" of advisors. It can also provide a formal group which interfaces with management on issues that affect women.

At AT&T, the Employee Counseling Service provides private and confidential counseling sessions for all employees and consults with management about special organizational issues. By presenting special programs on work place issues that impact women, it is able to affect the environment of the entire organization.

The issue of the adversarial climate is a complex one. The committee recognizes that styles of communication are very personal and individual, and that ONR is an organization with a military-oriented mission. However, its scientists and engineers are primarily civilians, and a style which alienates or intimidates employees is counterproductive. The committee does not believe that an adversarial, aggressive style of communicating is necessary to being effective in garnering dollars or managing a large research program, however hostile the environment. Being prepared and knowing how to make an effective, articulate argument are necessary. In the past two decades, many corporations have recognized the value of and adopted a more open, tolerant style and culture as they have learned how critical their human resources are to productivity.

Organizations that are successful in attracting and retaining women have an integrated approach to diversity. For example, at Apple the intent is to develop programs and interventions that support diversity at all phases of the employment cycle, including sourcing, recruiting, interviewing, orientation/acculturation, coaching, and development/promotion. To provide a more supportive climate for diversity, ONR will need to make changes throughout the organization. Changes to the work environment are perhaps the most difficult because they require changes in human behavior, but they are as critical as recruitment and hiring to long-term success.

ONR EFFORTS TO INCREASE DIVERSITY

The Office of Naval Research has initiated a number of activities in the past two years to increase the diversity of its S&E work force. In December 1994, the Deputy Chief of Naval Research chartered the Ad Hoc Diversity Working Group to examine this issue and propose appropriate action. The group recommended a two-pronged approach: (1) create an organizational climate that will provide opportunities for more meaningful participation by the targeted groups in all science and technology activities at ONR, and (2) implement initiatives to enhance recruitment and retention efforts for these groups.

In March 1995, the Chief of Naval Research disseminated an ONR Diversity Plan. He also created an ONR Diversity Committee under the STAB to study ONR diversity issues, demographics, and trends; to provide recommendations; and to measure and track progress.

ONR Diversity Plan

ONR leadership and the Diversity Committee are to be commended for their comprehensive Diversity Plan. It is thoughtful, comprehensive, and realistic in outlining the critical elements of a successful strategy to increase diversity. The committee is also impressed with the speed with which several of the elements of the plan have been implemented (see the section following on "New Initiatives"). As requested, this committee has a number of observations about individual elements of the plan.

- **Issue "All-Hands" Policy**. The committee agrees that issuing a policy sanctioned by senior staff is critical to the process of institutionalizing diversity as a value and of establishing diversity programs as essential to ONR's success. This approach can be strengthened by placing it in a strategic context with ONR's other goals to ensure that this policy is communicated clearly and consistently and that it has real impact on the daily lives of ONR employees.

- **Training**. (See the discussion under "New Initiatives.")

- **Internal Communications**. The committee applauds the idea of increasing internal ONR communications on all subjects. Issues related to diversity should be incorporated into standard ONR electronic newsletters or memos, however, rather than communicated through separate means, in order to reinforce the notion that diversity is central to ONR's culture and business.

- **External Communications**. The proposal to enlarge distribution of ONR press releases and external reports, for example, to improve access to target groups is an excellent demonstration of using existing practices to promote diversity. To ensure the success of this proposal, ONR must first understand the impact of its current external communication practices on hiring (e.g., How have candidates traditionally learned about ONR and opportunities for employment there?). When this is understood, similar methods can be employed with the target groups, paying careful attention to which media are selected. Professional associations representing the target groups can be very helpful in this process.

- **Mentoring Program**. (See the discussion under "New Initiatives.")

- **Departmental Plans**. Requiring each department to develop a diversity plan is an excellent goal that will be a critical component in institutionalizing diversity as a core value and common practice in ONR. Allowing for flexibility in implementation is important to cultivating a sense of ownership of diversity practices within each department and to creating initiatives that are appropriate for each group.

- **Advertising**. Extending the distribution of vacancy announcements to target populations is important but will not outweigh the factors that normally attract people to an organization, such as word-of-mouth information about working conditions and referrals when vacancies occur. In this context, more attention needs to be placed on creating an environment that supports a diverse population and is therefore attractive to a larger portion of the labor pool. Allowing potential candidates access to individuals who belong to the target population within ONR is an excellent way to introduce the organization to interested scientists and engineers.

- **Meetings**. Expanding ONR representation at conferences and conventions focusing on target groups is an excellent way to bridge the gap between ONR and scientists and engineers from target groups. Each event in which ONR participates, however, is only a milestone along the road of establishing an ongoing relationship with professional organizations that focus on minority or female scientists and engineers. ONR should assign a contact person for each such relevant organization and seek ways to support and communicate with the organization between events.

- **Networks**. Providing an Internet home page for members of target groups may be useful, but the plan does not provide a description of how it would work. A related issue to consider is how to encourage members of target groups to utilize information sources currently available to all.

- **Contacts**. The proposed plan of utilizing members of target groups with established relationships with ONR as points-of-contact in their communities is an excellent use of resources and should help strengthen the desired ongoing relationships.

- **IPAs**. The plan includes the establishment of eight new positions for IPAs in ONR departments to be filled by members of target groups. This approach addresses two barriers to increasing the representation of target groups in ONR: low turnover and lack of effective outreach to target groups. The proposed IPA positions may facilitate more aggressive recruitment within target groups. If these special IPA positions are filled, careful attention should be given to how these special hires are introduced into the organization.

New Initiatives

In slightly over a year, the ONR Diversity Committee has identified many useful ways to increase the representation of target groups. Included among them are new initiatives in four areas: diversity training, rotational assignments, mentoring, and recruitment and relocation bonuses.

Diversity Training

Training is helpful to ensure that common skills and knowledge are shared throughout the organization. However, without the opportunity for practical application of these skills and knowledge, it can become more of a theoretical concept and less of a value to the organization. Given the work force composition in ONR, generalized diversity training seems premature. Training needs to evolve as the organization changes. A more effective approach might be to provide training that is focused on specific areas that will directly impact the hiring, development, and retention of scientists and engineers in the targeted groups, such as interviewing skills, performance management, a system of rewards and recognition, and conflict management.

At the same time, making women and minorities more visible in the organization can be very beneficial. The "Environmental Consciousness" section of the diversity training plan suggests that departmental displays highlight principal investigators, their research, and pictures of diverse participants. This can be an excellent forum to demonstrate the contribution of scientists and engineers in the target groups. Presenting them along side their peers belonging to majority groups will avoid the impression of tokenism and will communicate the importance of scientists and engineers in the target groups.

The usefulness of producing a separate diversity brochure based on the mission and vision of ONR is not clear. A more effective way to communicate the importance of diversity in ONR might be to integrate diversity into the appropriate issues and practices described in existing media (new employee handbook, public affairs displays, newsletters, advertising campaigns, etc.).

Rotational Assignments

Providing rotational assignments is an excellent way to provide career development opportunities while also increasing collaboration and understanding among different naval organizations. As representation of the target groups increases, these scientists and engineers should be encouraged to participate.

Mentor Program

The pilot Mentor Program, announced in a February 1996 memorandum from the Chief of Naval Research, is designed to provide developmental opportunities to all ONR employees in scientific and engineering positions and to offer them access to, and prepare them for, future leadership roles. The basic design of this program appears well-suited to meet this goal.

The use of a "360 degree assessment tool," which provides feedback from an employee's managers, peers, and subordinates, is especially critical to the success of this program, and adequate attention needs to be paid to the administration of this process. Those providing feedback to the individual need to be encouraged to provide information that will help that individual succeed. The nature of this feedback is twofold: (1) to identify the person's existing behaviors and skills that are particularly important to succeeding in ONR and (2) to identify those behaviors and skills which may limit success. Although this may appear obvious, it is important to remind those who provide feedback of the importance of providing a balanced perspective.

The employee and the mentor should also be encouraged to dedicate sufficient time to the assessment to ensure that this information is used to its fullest. This tool provides a great deal of information, but the developmental plan that is generated from it must be considered a "living document" to reap the full benefits.

Providing mentoring to all employees is bound to improve both individual and organizational performance. To ensure success of this program, cultural differences need to be accounted for regarding such things as learning styles, conflict management, communication styles, and perceptions of authority. When members of the target group are paired with those in the majority group, there is an excellent opportunity for mutual cross-cultural learning.

It is also important, however, to recognize the need for members of individual targeted groups to spend time with one another as well and to provide a venue for these gatherings. This can be done in partnership with the appropriate professional associations and other naval organizations. To avoid exclusionary behavior, these meetings should be open to all members of the community.

The mentor program description includes among its objectives to "provide a tool to embrace diversity as a core value of ONR; to change the organizational culture to enhance the participation of minority, female, and disabled scientists and engineers; and to attract and retain qualified new people into ONR." This objective seems to exceed the scope of a mentor program, and the existing program description does not appear to include a tool or process that will meet this objective. If valuing diversity is to be one of the goals of the mentoring program, there should be greater clarity about how this is to be accomplished.

Recruitment and Relocation Bonuses

The ONR Diversity Committee has recommended the implementation of a recruitment and relocation bonus program to enhance recruitment of members of target groups in the S&E work force. The program would provide supplemental salary of up to 25 percent to new hires to ONR or to current federal employees in order to relocate. This committee questions the effectiveness of this approach to increasing diversity. The ONR Diversity Plan indicates that the dearth of members of the target groups in ONR is due to several factors, which include recruitment and hiring practices, low attrition, and the work environment. Providing recruitment and relocation bonuses are not likely to address these issues. At the same time, there is no evidence that the lack of hiring is due to an inability to meet the current market demands of female and minority candidates, which is the usual purpose of such bonus programs.

Also, the experience of organizations in industry and academia that have tried recruitment bonuses for new hires indicates that this practice can be counterproductive. In the interests of acquiring such bonuses for new employees, the selecting official may place more emphasis on a candidate's membership in a target group than on qualifications. Even in the cases where this is not true, these special hires may be perceived and, consequently, often treated as token hires who have not met the qualifications of the job. In either case, a tremendous disservice will have been done to both the scientist or engineer and to the organization as a whole.

There are, of course, exceptions, and managers should have the option of using recruitment or relocation bonuses where necessary to attract an outstanding candidate. However, bonuses to individual new hires should be rare.

Role of the ONR Diversity Committee

The ONR Diversity Committee is an important tool for ONR in its work to increase diversity, and the current committee is to be commended for its solid recommendations. This committee believes that the ONR Diversity Committee should be strengthened and given a position of greater prominence. Two ways to accomplish this are to add some senior executives to the committee and to have the Chief of Naval Research assume leadership in the position of Chair. This would send an unequivocal message to ONR scientists and engineers that the committee's functions are important to ONR's mission.

4

RECOMMENDATIONS: PART ONE

Creating a more diverse science and engineering work force at ONR is important to its future success, not only because of the demographic changes of the next several decades, but also because of the Navy's need to tap the very best minds from all sources for its research and development efforts. ONR has recognized that diversity is important, but a more focused approach than has been taken to date is necessary. The effort requires a public and repeated commitment from the very highest levels of ONR leadership. It requires strategic interventions in recruitment and hiring, as most ONR managers have already acknowledged, but also in the work environment for current and future employees. Top women and minorities will then be attracted to ONR as a place where the ability to direct an exciting technical program coincides with a supportive and productive environment.

Based on its findings, the committee makes two primary recommendations: to create specific targets for increasing diversity and to appoint an external group to monitor progress. In addition, a series of suggestions are provided for implementation of these recommendations in the areas of recruitment and hiring and employee development and climate. These suggestions focus primarily on process. The distinction between the primary recommendations and the suggestions for implementation is important; many well-intentioned changes to process can be made without any substantial change in the end result. The committee encourages ONR, in considering these recommendations, to focus on the primary goal of hiring and keeping more minorities and women and not to be content with simply making good faith efforts.

Although the proportion of underrepresented scientists and engineers in the general population is small, such individuals do exist, and their numbers are increasing. Other organizations have succeeded in increasing diversity, and much can be learned from their experiences. ONR has its finger on the pulse of American research and development through its almost 5,400 principal investigators, and they can and should be recruited into this effort. The committee urges ONR to approach this challenge as it would any other: to do the hard thinking, to develop the plan, to dedicate the resources, and to follow through until it is completed. There is no question that an organization with the history, credentials, and reputation for excellence of the Office of Naval Research can accomplish this task.

PRIMARY RECOMMENDATIONS

RECOMMENDATION #1
The Chief of Naval Research should assume the responsibility to develop specific, numerical targets for the hiring of women, minorities, and persons with disabilities into science and engineering positions at ONR. These targets should be based on a periodic assessment of the underutilization of qualified individuals from these groups, using data from national pools.

The effort should include communicating regularly to ONR personnel the importance of diversity and of meeting these targets. The targets should cover senior executive positions as well as those at the GS13–15 levels. They should be based on ONR's staffing needs, the current pool of qualified underrepresented personnel available in the relevant fields, projected turnover, and other budget and personnel constraints. Managers would then be held accountable for meeting appropriate targets in their areas based on demonstrated underutilization, and their success would be reflected in performance reviews, salary increases, cash awards, and promotions.

Targets should be reviewed periodically and readjusted based on changes in the ONR work force and the most recent available data on underutilization from the national pools. The pools of qualified personnel can be determined by special tabulations from the Survey of Doctorate Recipients and the National Survey of College Graduates tailored to ONR's work force needs, as has been done with the national pools of scientists and engineers described in this report.

Target levels for increased diversity should be developed in conjunction with formal plans for future personnel needs. Sometimes called "succession planning" or "people planning," each department head should develop a personnel management plan to include identifying who might succeed the current incumbents in the department, how to trade or share employees among divisions, and how to bring more minorities and women with the necessary background and interests to work at ONR.

RECOMMENDATION #2
The Chief of Naval Research should appoint an external committee composed of individuals who are experienced in the management of science and engineering and sensitive to the issues of diversity to assist ONR in achieving its diversity goals. Reporting to the Chief of Naval Research, the committee would meet periodically to review the targets ONR has set and to evaluate progress against those goals.

The external committee would be different from the Boards of Visitors, which ONR convenes to evaluate divisions and research programs, in that it would be charged with the issue of diversity only and would review the Science and Technology Directorate at ONR. The committee would consist of science and engineering managers from business, academia, and government, and especially from organizations with a strong record of accomplishment in diversity. It would include individuals knowledgeable about ONR's mission and would have representation from underrepresented groups.

SUGGESTIONS FOR IMPLEMENTATION

The recommendations in this section outline specific steps which ONR can take to help implement the diversity targets described above. They are not ends in themselves but means to an end. This section is not intended to include an exhaustive list but to provide examples of the types of initiatives that can help ONR meet its needs. The suggestions cover two areas: (1) recruitment and hiring and (2) the work environment; both are critical to creating and sustaining diversity in the work force.

> RECOMMENDATION #3
> **ONR should expand its recruitment efforts and improve the hiring process to increase the likelihood that members of the target groups will learn about positions at ONR, will apply, and will be given serious consideration.**

Advertising and Recruitment

- **Educate the at-large scientific and engineering community, and especially underrepresented scientists and engineers, about the advantages of working for ONR.** Provide information on opportunities to carry out research and engage in joint activities with universities or industrial organizations. Discuss ONR's commitment to diversity and the changes under way to create a work environment supportive of women and minorities. Where possible, send women and minority scientists and engineers from ONR and NRL to speak in universities and at professional meetings as role models for students and faculty and to reward those employees for their extra efforts.

- **Develop contacts with professional organizations of underrepresented scientists and engineers.** Establish ongoing relationships with individuals, foundations, and societies that truly have expertise in where to locate women and ethnic minority scientists and engineers in each field. Advertising in journals is not enough (and indeed may not be at all effective) when seeking scientists and engineers from the targeted populations. Use the minorities and women already on the staff at ONR to identify places in which to advertise and especially people to contact.

- **Advertise nationally and recruit outside the Navy.** The work of ONR is national in scope, and its S&E work force should be drawn from a national pool. Underrepresented Ph.D. scientists and engineers are located overwhelmingly (75 percent) in universities and industry, not in the federal government. They are also located across the country, not just in the Washington, D.C. metropolitan region. Recruitment actions that are limited to individuals in that region or to those who already work in the Navy or in the federal government are less likely to accomplish ONR's diversity goals. If experience in the Navy is deemed important, ONR should create ways to provide that experience after hiring. Giving new employees Navy assignments at the beginning of their tenure or putting them through an orientation program, supplemented by visits to Navy installations, are possible ways to accomplish this objective.

- **Recruit women and minorities more actively from within the Navy's associated laboratories and other facilities.** This approach allows ONR to hire personnel with Navy experience, to the extent that underrepresented scientists and engineers are employed in these facilities. There are, for example, 170 female scientists and engineers at NRL, of whom 10 are at the GS15 or senior executive level. There are also 44 scientists and engineers at NRL who are members of minority groups, with 5 at the GS15 or senior executive level. In addition, individuals who receive Postdoctoral Research, Young Investigator, or Women Science Scholar awards through ONR's corporate programs are strong potential candidates for program officer positions, given their postdoctoral education and experience with ONR (see Recommendation #6 in Chapter 6).

- **Encourage and assist the Navy laboratories to recruit more actively from targeted groups.** Many of the same mechanisms recommended here for ONR could be used by the laboratories. This will help improve the diversity in other parts of the Navy's R&D enterprise and increase the potential pool of future ONR employees. ONR should assume the obligation of assisting Navy laboratories in general, and NRL in particular, in increasing diversity, which, in turn, would be to the interest of all parties. Taking on such an external commitment will not only be important for the Navy, but it will focus ONR's attention on its own diversity needs and options.

- **Make particular efforts to recruit women and ethnic minority scientists actively for senior executive positions.** If diversification occurs at the top level (senior executive positions), it will send a signal to the lower ranks and to the outside community that ONR is committed to diversity. The IPA vehicle can be used to bring individuals into ONR for short-term assignments, similar to what NSF does in its Rotational Programs for Scientists, Engineers, and Educators. ONR could also recruit faculty members to spend their sabbaticals at headquarters or to develop 3–5 year partnership programs with academia or industry. In the same vein, the IPA program can be used to give senior executives opportunities to work in another organization outside the government. It may be useful to arrange assignments in a Navy laboratory for one to two years, especially for new ONR employees who lack adequate Navy experience.

Hiring Process

- **Broaden job descriptions and selection criteria to eliminate unnecessary restrictions.** Job descriptions that are very narrowly defined often discourage candidates, especially minorities and women, who are well qualified but may be lacking in one specific but less critical area that can be acquired later. Such job descriptions may also send the message that the desired candidate has already been identified and that the application process is a waste of the potential applicant's time. Similarly, applicants with high potential without the exact requirements in the job description may be eliminated early in the selection process without having been given an opportunity to demonstrate their ability. One way to avoid this loss is to invite a larger number of minority and female candidates to participate in interviews.

- **Broaden and diversify the decision-making process.** A critical group in ONR's current hiring process is the rating and review panel. All rating and review panels should include women and minorities either drawn from the ONR work force or invited from elsewhere in the Navy, industry, or academia. The Selecting Official should interview an appreciable number of qualified candidates, including members of underrepresented groups, so long as "token" interviews are avoided. Finalists should make presentations to a broader group of ONR scientists and engineers, not just to the STAB, in order to obtain feedback from a larger group, to promote a less adversarial setting for interviews, and to establish broader knowledge about the hiring process. Finally, before approving the hiring of any individual, the Deputy Chief of Naval Research should be provided with evidence that a sufficient effort has been made at each step in the hiring process to recruit and hire a member of the target group. Some of that evidence would consist of documenting actions recommended here.

- **Set aside modest discretionary funds to assist in the hiring of candidates with especially high potential.** Although the committee does not recommend the regular use of hiring bonuses, discretionary funds might be used in rare cases to provide recruitment or relocation bonuses or to meet other needs of highly recruited candidates, especially for senior executive positions. These funds would enable ONR to compete more effectively for outstanding female and minority applicants who are also being recruited elsewhere.

> RECOMMENDATION #4
> **ONR should improve the work environment to increase productivity, enhance employee development, and establish ONR as a place where women and minorities want to work.**

Work Environment

- **Promote diversity on a continuous basis in ways which are visible to the entire work force.** This includes publishing diversity targets and regular discussion of diversity in oral and written communications from the Chief and Deputy Chief of Naval Research and the department heads. It could also include special recognition for managers who have made significant strides in increasing diversity in their units, as well as showcasing the accomplishments of female and minority scientists and engineers. Another way to promote diversity is to include more women and minorities on the Boards of Visitors and other oversight bodies.

- **Create mechanisms for periodically assessing employees' needs and concerns.** Conduct a Work Place Climate Survey of employees every two years, as is done, for example, at the Xerox Corporation, and hold managers responsible for the work climate in their unit. At Xerox, managers' performance reviews and salary increases are affected by the survey's results. In addition, ONR leadership should spend more time talking to the staff. Several program officers, in their interviews, recommended that the top management talk more often with those below them. This might be done through informal breakfast meetings or coffee with groups of program officers. The purpose of the meetings would be for senior managers to listen to program officers' ideas and concerns, to answer questions, and to get their feedback on ONR policies and directions.

- **Institute changes that will help all employees be more productive by providing greater flexibility for the individual.** Such changes might include the creation of flextime and flexiplace policies, to the extent that the technology will support them. It could also include increased options for part-time work or job sharing. Many employers also help to accommodate for the extra demands placed on single parents or dual-career couples through programs of parental or family leave.

- **Foster a less adversarial or confrontational style of communicating, especially in interviews with job candidates.** As the work force at ONR becomes more diverse, the committee expects that a natural consequence will be the evolution to a less adversarial style and one that is more conducive to variety and creativity. In the meantime, this process could be accelerated by some amelioration of the current style. An adversarial approach is especially counterproductive, in the committee's view, in the hiring and interview process. Once hired, program officers who need help in making forceful and articulate briefings or budget presentations should receive special training in those skills. Many corporations and federal agencies, for example, prepare congressional witnesses by holding friendly "murder boards" where high-level managers, including the director, try to anticipate the "hostile" questions and help the individuals formulate their answers.

- **Collect more detailed information on employee departures and attrition rates.** Although ONR does track employee departures, including the general reason for leaving, additional information should be collected. Data on attrition rates should be calculated and reported to ONR leadership annually, including data by race, gender, and disability. Also, confidential exit interviews should be conducted with all departing employees to determine if there are aspects of the job or work environment that could be improved.

Employee Development

- **Create a "personal development plan" for each staff member.** This plan would be worked out between the staff member and management. It should cover at least five years and address (1) training needs, (2) plans to correct perceived deficiencies in performance, (3) possible rotational assignments in and out of ONR, and (4) potential promotion paths, where available. In the case of senior executives, the plans should be used to promote new opportunities and greater turnover through assignments to other Navy facilities or government agencies or IPA's to work in industry or universities, as well as to identify incentives for early retirement, when appropriate. In the case of employees with master's degrees, the plans might include leave or a reduced work load to pursue a Ph.D.

- **Provide opportunities for advancement for women and underrepresented minorities**. Given the absence of women at the senior executive level, particular effort should be made to identify experienced female program officers for promotion, especially into the senior executive ranks. Other opportunities for advancement include appointing women and underrepresented minorities as acting heads of divisions or departments, inviting them to make formal presentations and briefings, making the Research Opportunities for Program Officers (ROPO) program available to applied scientists, facilitating both permanent and temporary internal job rotation, and encouraging the pursuit of advanced education and training. To the extent that detailees and IPA's are seen as likely candidates for ONR permanent positions, they should be given opportunities for greater visibility and responsibility as well.

- **Establish a formal coaching program.** The committee prefers this term to "mentoring" because the latter sometimes implies one high-level person grooming an individual to fill a specific career track. "Coaching" is usually done by people at various levels—generally not by one's supervisor—and does not need to have a specific career goal in mind. Coaching works best when it is informal. However, not all employees have access to individuals willing to coach them, and not everyone is good at coaching. Training on how to ask for help and how to be a good coach can be helpful. In the meantime, though, a formal pairing of employees with "coaches" other than their supervisors can serve as a start on more natural, informal contacts. This should help improve communication both upwards and downwards and reduce the sense of isolation many program officers mentioned in their interviews. ONR's pilot Mentor Program is a good start in this direction.

- **Provide task-specific training for program officers as well as for senior executives.** The usual purpose of diversity training is to teach employees how to get along better and work more productively with a diverse population of colleagues. In the case of ONR, however, there may not be enough racial and gender diversity yet to make this useful. Also, if done poorly, such training can be counterproductive, setting groups off against each other and reinforcing existing prejudices.

A training program that evolves as the organization becomes more diverse and that is task-specific, however, can be very helpful. Managers responsible for recruitment and hiring, for example, could attend workshops on how to advertise for and recruit a more diverse applicant pool and on interview techniques. Program officers could receive training in assertiveness, negotiation, and public speaking in order to increase their effectiveness in program briefings and budget presentations.

- **Establish performance reviews that reward supervisors for gains made in diversity**. Use success in this area as one factor in performance ratings and in awarding salary increases and cash bonuses. "Success" would include gains in hiring and promotion within the ONR S&E work force as well as in efforts to increase the pool of potential future ONR employees. Individuals should be asked to prepare a separate, one-page statement of what they have done in the previous year to help enhance diversity in the ONR S&E work force.

PART TWO

USING ONR'S CORPORATE PROGRAMS TO ENHANCE DIVERSITY

5

OBSERVATIONS ON ONR'S CORPORATE PROGRAMS

OVERVIEW

Like most other mission agencies, the Office of Naval Research administers a substantial portfolio of multidisciplinary research and education programs that support its mission as a whole. The purpose of these "corporate programs," as they are called, is to increase the number of engineers and scientists engaged in technological efforts of concern to national defense and specifically to the Navy. ONR recognizes that it is not enough to fund current research and development to support the Navy's mission; funds must be invested to educate and sustain the future pool of scientists and engineers to maintain the capability to respond to future needs.

The primary purpose of the corporate programs is not to increase the diversity of the science and engineering work force. However, as discussed in Chapter 1, the U.S. population as a whole (and school-age children in particular) is becoming more racially and ethnically diverse. Federal education and research programs will need to draw more aggressively from all groups within this population to identify the best talent. Because they reach individuals from high school through postdoctoral study, the corporate programs are important tools that ONR uses to help identify students interested in science and engineering, to encourage them through their academic careers, and to help establish them as emerging researchers. With some thought and attention to where the populations of non-white students and faculty are and how to reach them, the same tools can be used to help increase the diversity of the pool of scientists and engineers.

With this overall goal in mind, ONR requested the NRC to review its corporate programs and to recommend ways they could be improved to accomplish two objectives:

- To help increase the diversity of the nation's overall pool of scientists and engineers working in areas of relevance to the Navy

- To help increase the diversity of the scientists and engineers working at ONR

All of the corporate programs have the potential to help increase diversity of the overall pool, but only a few—those at the postdoctoral level—might realistically be used as tools to bring a more diverse work force into ONR at any time in the near future.

The committee was also asked to comment on the effectiveness of the corporate programs in general, where appropriate. It has attempted to do so, but it has made no attempt to analyze or evaluate each of these programs in depth. That kind of program evaluation is more appropriate for a committee or organization with different expertise and a single, focused charge. Although this committee has offered some comments on how to strengthen the entire portfolio of corporate programs, the primary interest here is to respond to the specific question of how these programs might be enhanced to contribute more actively to diversity.

The programs that, in the committee's judgment, are relevant to this discussion consist of the 16 educational and research programs that support high school students, undergraduates, graduate students, and young postdoctoral researchers. These programs totaled $35 million in FY96, only $12 million of which came directly from ONR funds. The remaining $23 million consisted of funds from the Office of the Secretary of Defense (OSD) that were administered by ONR. The $11 million used to support Historically Black Colleges and Universities (HBCUs) and Minority Institutions (MIs)[1] came from both ONR and OSD; these programs help to meet the Department of Defense (DoD) goal that 5 percent of each agency's research funds be set aside for programs for these institutions.

The 16 programs are listed below with their FY96 budget allocations.

High School Programs **$1.4M**

- Science and Engineering Apprenticeship Program (SEAP)
- High School Apprenticeship Program
- Naval Science Awards

Programs for Undergraduate Students **$1.1M**

- Augmentation Awards for Science and Engineering Research Training (AASERT)

Programs for Graduate Students **$13.5M**

- Augmentation Awards for Science and Engineering Research Training (AASERT)
- National Defense Science and Engineering Graduate (NDSEG) Fellowships
- ONR Graduate Fellowships (being phased out)

Programs for Postdoctoral Researchers **$8.0M**

- Postdoctoral Research Program
- Young Investigator Program
- Summer Faculty Research Program
- Women Science Scholars
- Ocean Science Educators Awards

Programs at HBCUs and MIs **$11.0M**

- HBCU/MI Science and Engineering Education
- HBCU Graduate Fellowships
- HBCU Engineering Faculty Fellows
- HBCU/MI DoD Infrastructure Support

TOTAL **$35.0M**

---

[1]The list of Historically Black Colleges and Universities and Minority Institutions can be obtained from the U.S. Department of Education, National Center for Education Statistics.

THE INDIVIDUAL PROGRAMS

As part of its analysis of the corporate programs, the committee reviewed promotional materials, budget information, data on the numbers and type of participants, and both qualitative and quantitative evaluations, where available. It also discussed the goals and objectives of the programs with ONR managers. Below are brief descriptions of the major programs, along with observations that may be helpful in enhancing these programs' effectiveness for the Navy's mission and for increasing diversity in the science and engineering pool.

High School Programs

Programs to interest high school students in science and technology and to give them hands-on experience in research contribute to sustaining the pool of future scientists and engineers, and especially to enlarging that pool to include more minorities and women. It is at the high school level, or earlier, that many promising students are lost to science, math, and other technical fields. ONR offers several apprenticeship programs for high school students as well as a small one for junior high students.

The largest program is the Science and Engineering Apprenticeship Program (SEAP), which is funded by the OSD. SEAP gives 600 students in the Washington, D.C. area an opportunity to assist in ongoing research and development work in Navy and other DoD laboratories during the summer. Based on a 1994 survey of 1,064 former participants conducted for ONR by the Academy for Educational Development, 70 percent still indicated interest in science and engineering as a career. Forty-five percent of the program participants were female, 7 percent African American, and another 5 percent listed as "other." The percent of females is encouraging, but the percent of African American students is not, especially since over 80 percent of the high school students in the District of Columbia are African American.

By contrast, ONR's High School Apprenticeship Program, which sends high school students into the laboratories of ONR principal investigators at U.S. universities, has a high minority participation (80 percent in 1994). This is most likely due to the fact that the program was formerly targeted only for minorities, and the contacts developed during that time continue to exist. ONR staff report, however, that the number of minority participants is declining now that it is no longer targeted to minorities. This program is a good example of the success which can be accomplished with a concerted effort to recruit students from racially diverse high schools. It is dependent, however, on the initiative of individual ONR program officers and their principal investigators to identify potential participants.

Programs for Undergraduate Students

The only program for undergraduate students, other than those at HBCUs/MIs, is the Augmentation Awards for Science and Engineering Research Training (AASERT). This program, open to U.S. citizens only, allows ONR principal investigators to fund undergraduates and some high school students as research assistants on their ONR grant or contract in order to stimulate student interest in research and to provide actual experience in original investigation. A similar program is available to graduate students. AASERT was begun in 1991 in response to congressional concern about a potential shortage of Ph.D. scientists and engineers as well as a desire to increase the number of U.S. citizens holding research assistantships. The program allows ONR investigators to supplement their grants with additional research assistants who would not otherwise be funded by that grant. In 1994, 428 graduate and undergraduate students were supported—239 of them by direct ONR funds. Thirty-five percent of those students were women and 3 percent were minorities, lower than the 44 percent of female baccalaureate graduates in science and engineering and the 11 percent of minority graduates reported in 1991 (NSF 1994, pp. 54, 57).

Representing $9M of the $35M spent annually in educational programs, AASERT is essentially a way of supplementing already funded faculty research grants, but with some eligibility restrictions. The presumed shortages of Ph.D.s predicted in the late 1980s and early 1990s do not seem to have materialized, but the undergraduate portion of the program, at least, is very much in line with the need to broaden the experience of undergraduates in science and engineering. It is unfortunate that a program of this size and potential value has not attracted a more diverse population.

Programs for Graduate Students

In addition to the AASERT program described above, ONR's main support at the graduate level is through the National Defense Science and Engineering Graduate (NDSEG) Fellowship program. (The related ONR Graduate Fellowship program is being phased out.) The NDSEG program has been in existence since 1989. It funds about 90 new fellowships annually, providing stipend and tuition support for doctoral students in ONR-related disciplines. The program currently is about 30 percent female and 6 percent minority. That latter number has declined in the past year coinciding with the removal of a requirement that 10 percent of the awards be targeted for minorities.

ONR participation in the NDSEG fellowship program is too recent to yield significant results in the form of Ph.D. production, but some data are available on the earlier ONR Graduate Fellowship program (started in 1980). Based on data collected by the American Society for Engineering Education (ASEE), ONR made 594 fellowship awards between 1982 and 1994. By the spring of 1995, at least 245 students had received the Ph.D. Forty-seven percent took jobs in industry and 24 percent in universities, with another 14 percent pursuing postdoctoral research. No data are available on degree completion or employment by race or gender.

The graduate fellowship program is a relatively generous one, with stipends of $16,500 to $18,500, and it is not surprising that most students finish the three-year program. However, more significant are whether they complete their degrees and what they do upon graduation. Also, unlike the programs for young faculty members, these students do not appear to have any contact with ONR once the award is made since the fellowships are portable and are not connected with a Navy organization or other ONR research activity. This pool of doctoral students, in the committee's opinion, represents a largely untapped resource for ONR as it expands the pool of potential investigators and employees.

Programs for Postdoctoral Researchers

Programs for postdoctoral researchers offer the most promise for identifying women and minorities as possible future investigators and program officers for ONR. The Postdoctoral Research program brings promising scientists and engineers into a Navy laboratory to work on an area of interest. Of the 213 participants in this program between 1990 and 1996, 40 percent continued to work in a Navy laboratory after completing the fellowship. Twenty percent went to academia, and another 20 percent went to industry.

The Young Investigator program funds young researchers with $100,000 for three years to do Navy-related work on their home campuses. In 1996, 34 individuals out of 416 applicants were funded. Many of these individuals go on to receive ONR research grants. No systematic demographic data are available.

The Summer Faculty Research program funds about 100 faculty members per year, about half of whom are from HBCUs and MIs. Participants spend 10 weeks over the summer working at a Navy laboratory. Begun in 1979, this program has funded 2,670 faculty members to date, many of whom return in subsequent years.

The participants in all of these programs already have a strong connection to ONR through their colleagues in the Navy laboratories or through the ONR program officer who monitors their progress. Thus they have a natural entree into ONR and the Navy's research and development efforts. No data on the race or gender of the participants are collected in any of these programs, so there is no way to know how many women or minorities receive these awards. In addition, the committee's impression is that candidates for these awards are often identified informally through suggestions from colleagues. This can limit the number of minorities and women who are considered. If special efforts were made to recruit members of underrepresented groups into these programs, ONR would have a rich resource on which to draw for possible new program officers.

Unlike the above programs, the Women Science Scholars program does not connect participants to the work of the Navy or of ONR in any specific way. Instead, it provides a year of support to enhance the careers of female scientists and engineers generally. In conjunction with area research facilities, eight women are selected each year to work at the Bunting Institute of Radcliffe College. The specific focus of the program is on individuals who are at "critical career points" in their advancement as scholars (e.g., close to tenure, reentering the work force after time away, moving from a liberal arts to a research-intensive institution, or changing fields). Applications for this program come primarily from the Boston area, most likely because of the difficulties of relocating for one year. Also, applications are overwhelmingly in the life sciences, with few from the physical sciences or engineering, which are the dominant areas of research and development at ONR. Program staff are working to ensure greater breadth in disciplines, but the problem of temporary relocation from other parts of the country remains a difficult one.

The committee noted that participants in this program are not currently required to do research in an area of interest to the Navy. This means that they are not connected to ONR or to a Navy laboratory the way other postdoctoral researchers are. The Navy may thus not be getting as direct a benefit from this program as it might. Equally important, ONR may be missing an opportunity to recruit program officers or at least principal investigators or laboratory employees from a potentially valuable pool of experienced female scientists and engineers. Some modification of this program to tie it more closely to the research agenda of ONR and to a broader geographic area may be in order.

Programs at HBCUs and MIs

Educational programs at HBCUs and MIs consist primarily of undergraduate and graduate student fellowships. Recipients of all of these awards must be U.S. citizens. The Science and Engineering Education program provides five-year grants to HBCUs and MIs to increase the number and quality of baccalaureate students in those fields, with a long-term goal of preparing them for graduate study. Grant funds typically go toward scholarships, bridging programs, faculty exchanges, and similar mechanisms.

The other major HBCU/MI program is the DoD Infrastructure Support Program, which provides support to the institutions directly for curricular enhancements, faculty exchanges, equipment acquisition, and other activities to build the infrastructure necessary to teach and conduct science and engineering. Begun in 1993, the program provides about $4 million annually to about eight institutions.

The HBCU Graduate Fellowship program provides up to four years of full support for doctoral students who are graduates of HBCUs. Graduate fellowships can be taken to any institution. Since the initial awards in 1992, 16 students have been selected for the program, but so far six have left the program without completing their graduate studies. The number of applications for the program is low considering the promise of four years of support.

Although it is too soon to evaluate the success of the program, some of these early indicators are troubling. Recruiting more effectively, identifying a strong applicant pool, and possibly providing special support for students in their transition to graduate school may help strengthen the program.

A similar program supports graduate students who agree to serve as engineering faculty members at an HBCU after completing their degrees. Fairly limited in scope, the HBCU Engineering Faculty Fellows program provides three years of support to three new students annually. Since its inception in 1992, all of the recipients have been members of minority groups, and two-thirds have been female.

THE ENTIRE PORTFOLIO

The committee was generally impressed with the breadth of educational programs ONR administers and with the dedication of its program staff. While these programs do not directly address ONR's primary mission of research and development, they are important to achieving that mission by helping to develop the scientists and engineers who can assist in solving the Navy's technical challenges in the future.

ONR's corporate programs are relatively young, typically beginning in the early 1980s as fellowship programs. The HBCU/MI programs, especially, are only a few years old. The committee's impression is that these programs evolved historically, as the Navy perceived a need or as Congress mandated a particular activity, and, while each is worthwhile in its own right, they do not necessarily form as coherent a set of programs as they might. Some of the programs are extremely small, consisting of fewer than 10 awards per year and budgets of $100,000 to $300,000. Some programs are restricted to specific geographic areas like the Washington, D.C. metropolitan area, which is close to ONR headquarters and to several Navy facilities. Others are limited to the HBCUs and a smaller number of MIs.

None of these characteristics seem consistent with the nature of ONR as a national agency supporting education and research. They seem, rather, to be characteristics of programs which grew up independently, as such programs often do, each with its own history and clientele. They need to be related in a coherent structure to a single set of goals, and they should reach science and engineering talent nationally.

Because the programs are fairly young and because data collection and evaluation have been uneven, there is not yet an adequate track record, in many cases, by which to measure the long-term success of these programs. Such measurement will be necessary if individual programs are to be placed in the context of overall goals. If one of those goals is to help increase the diversity of the national science and engineering pool, data collection and evaluation must be improved. At a minimum, data on the gender and race of all applicants and awardees are required, along with tracking mechanisms to find out where program participants go.

6

RECOMMENDATIONS: PART TWO

The committee's recommendations on ONR's corporate programs are of two types: primary recommendations and suggestions for implementation. The two primary recommendations—to create a single vision for the programs and to provide a continuum of educational support—speak to the entire portfolio of corporate programs. The suggestions for implementation address specific programs or aspects of the programs that will help ensure that they meet the primary objectives.

PRIMARY RECOMMENDATIONS

> RECOMMENDATION #1
> **ONR should create a single, coherent vision for the corporate programs, tied more closely to the mission of ONR.**

ONR needs a clear, integrated vision for what it wants to accomplish with these programs, including ways to use them to increase the number of women and minorities in science and engineering. It should then design a plan to make such changes as may be required. This plan should tie each program clearly to the research and development interests of the Navy, especially at the more advanced (postdoctoral) levels where ONR recruits.

Such a vision will increase each program's individual effectiveness by allowing it to build on the others. It will also create a framework for making decisions about when and whether to add or phase out programs and can provide a basis for ONR to resist ad hoc requests for new programs targeted at less essential needs. The committee's impression is that there are too many programs, some very small, and some that have evolved from various unrelated initiatives. This recommendation is not limited only to programs that ONR funds; the agency should exercise a leadership role, encouraging the Office of the Secretary of Defense to align all such programs with the long-term goals of the Department in the area of scientific and engineering personnel.

To integrate these programs more fully into the Navy's mission, ONR should create incentives for program officers who fund the core research and development program to become more involved in the corporate programs. This might include visits to fellowship recipients at the undergraduate and gradate levels on college campuses and increased involvement in high school and summer programs, as well as regular meetings with young investigators, postdoctoral researchers, and female scholars. This outreach will also increase the visibility and presence of ONR on college campuses. Performance reviews should reflect individual program officers' success at meeting these goals.

RECOMMENDATION #2
ONR should realign its corporate programs to provide a continuum of educational opportunities from high school through postdoctoral study.

For all potential students and young researchers, but for underrepresented groups especially, a continuum of support, from one educational level to the next and into one's early career, is important. The committee's impression is that the efforts made by ONR at the high school level, for example, are not followed through with any substantial programs at the undergraduate level. Similarly, graduate students with ONR or NDSEG fellowships are not necessarily connected in a systematic way to postdoctoral or other research opportunities. ONR needs to be visible and active at each level in the educational process. The effectiveness of funds invested at one stage can be amplified by channeling some of the same students through multiple levels of ONR programs. In addition, academic year programs can be leveraged by providing recipients with follow-up internships or summer jobs at Navy facilities. At the very least, students who receive ONR funding should be made aware more systematically.

The greatest break in support seems to come at the undergraduate level. Except for awards to the HBCUs/MIs and the occasional undergraduate who participates in a principal investigator's research grant under AASERT, ONR funds nothing at this level. The committee recommends that ONR consider creating undergraduate scholarships or a summer intern program at selected universities or Navy laboratories.

SUGGESTIONS FOR IMPLEMENTATION

RECOMMENDATION #3
ONR should use broader criteria for recruitment and selection.

As with employment decisions, it is easy in fellowship or research programs to fall back on two or three quantitative measures of potential success such as grades and test scores or, at a more advanced level, to rely solely on the opinions of trusted colleagues. Such methods of recruiting and selecting participants for these programs can, however, eliminate strong candidates from consideration. In the first instance, those eliminated candidates may not have the highest quantitative measures of success, even though they show great potential or evidence of success in other ventures. In the second instance, they are seldom known to a group of peers that consists primarily of white males. Several of the recommendations about recruitment and hiring for ONR's work force in Chapter 4 can be adapted for use in these programs.

RECOMMENDATION #4
ONR should extend support for minority students and faculty beyond the Historically Black Colleges and Universities (HBCUs) and Minority Institutions (MIs).

Like many federal agencies, ONR has an extensive effort supporting the HBCUs and MIs, tied closely to, but going beyond, the 5 percent required "set aside." These are good programs, but they reach only a small percent of the minority students in science and engineering, especially at the graduate level. Large urban institutions, and others with especially strong track records of success in educating minority Ph.D.s, should also be targeted for funding.

If ONR is to use its corporate programs to increase diversity in its work force and, more specifically, to expand the pool of potential employees, it will need to target a much wider audience of minority graduate students and researchers. One effective way to do this is through predoctoral and postdoctoral training grants to universities that have a history of, or creative mechanisms for, educating scientists who are underrepresented in ONR. The NIH has now created comparable initiatives to try and increase the pool of underrepresented minorities at the doctoral level who will have the best education possible in the area of biomedical sciences.

RECOMMENDATION #5
ONR should collect systematic data on program participants to allow for ongoing program evaluation.

Data collection and evaluation are not just good practice; they are now required under the Government Performance and Results Act (GPRA) of 1993. Although data collection has occurred in some programs, it has been uneven in others. Data on the gender and race of participants will allow managers to identify members of underrepresented groups for potential future recruitment. It will also allow some evaluation of the effectiveness of each program in helping to increase the diversity of the pool from which ONR recruits. With these data as sparse as they currently are, there is no way to assess the effectiveness of these programs in reaching a diverse clientele. While there are privacy rights to be respected, the committee believes it is possible to collect these data by a process that is separate from, but parallel to, the application process for each award so that individual applicants will be assured that selection decisions are not influenced by personal characteristics.

Equally important is a system to track the progress of program participants. ONR should create, or require its contractors to create, a system by which participants can be followed while they are in the program. They should also create a mechanism for systematic follow-up with a sample of program participants after they complete a program. In addition to providing necessary data for evaluation, tracking program graduates into college, graduate school, or their first careers will allow ONR to create a potential clientele for future research programs as well as job opportunities. It is the only way to know if individual programs—well designed in their own right—are truly contributing to the agency's overall mission.

> **RECOMMENDATION #6**
> **ONR should use the programs for postdoctoral researchers to recruit more aggressively for potential ONR employment.**

Individuals with Postdoctoral, Young Investigator, or Women Science Scholars awards are natural candidates for ONR program officer positions, if not immediately following their award, at least reasonably soon thereafter. ONR should use the same advertising and recruitment techniques for this group as is recommended here to increase the diversity of its current work force. While on tenure, these individuals could be brought to ONR headquarters to attend an orientation about what ONR does, to meet program officers face to face, and to tour some Navy facilities. They should also be tracked after completion of their awards as possible future job prospects and as members of boards of visitors or selection panels.

In the case of the Women Science Scholars, the program should be redesigned to fit more closely into the mission of ONR. Specifically, the fields designated for research should better match fields of interest to ONR, and the relevant program officer(s) should be involved more directly in making these awards and tracking the progress of these scientists and engineers. This kind of program has the potential of yielding some strong employees for ONR or other Navy organizations, but only if the scholars' research fields are of strategic value to the Navy.

REFERENCES

Adler, Nancy J.
1986 *International Dimensions of Organizational Behavior*. Boston, MA: Kent Publishing Company.

Cox, Taylor H., and Stacy Blake
1991 "Managing cultural diversity: implications for organizational competitiveness." *Academy of Management Executive* 5 (3): 45–56.

Hoffman, L. Richard, and Norman R. F. Maier
1961 "Quality and Acceptance of Problem Solutions by Members of Homogeneous and Heterogeneous Groups." *Journal of Abnormal and Social Psychology* 62 (2): 401–7.

Jackson, Susan E.
1989 "Team Composition in Organizational Settings: Issues in Managing a Diverse Work Force." In *Group Process and Productivity*, edited by J. Simpson, S. Warchel, and W. Wood. Beverly Hills, CA: Sage Publications.

Moss Kanter, Rosabeth
1983 "The Change Masters." New York: Simon and Schuster.

National Research Council (**NRC**)
1990–
1995 Survey of Earned Doctorates. Washington, DC: National Academy Press.

1993 Survey of Doctorate Recipients. Washington, DC: National Academy Press.

National Science Foundation (**NSF**)
1993 National Survey of College Graduates. Arlington, VA: NSF.

1994 *Women, Minorities, and Persons with Disabilities in Science and Engineering: 1994*. NSF 94–333. Arlington, VA: NSF.

Nemeth, Charlan Jeanne
1986 "Differential Contributions of Majority and Minority Influence." *Psychological Review* 93 (1): 23–32.

Whitten, Barbara L.
1996 "What Physics Is Fundamental Physics? Feminist Implications of Physicists' Debate over the Superconducting Supercollider." *National Women's Studies Association Journal* 8(2): 1–16.

APPENDIX A

DATA TABLES

TABLE A-1.1 Number of Employed S&E Doctorates (1960–89 graduates), by Field of Doctorate and Demographic Characteristics, 1993

Demographic Characteristics	Total		Biological Sciences		Engineering		Mathematics & Computer Sciences		Physical Sciences		Psychology	
	#	%	#	%	#	%	#	%	#	%	#	%
Total	228,442	100.0	68,432	100.0	49,854	100.0	19,306	100.0	68,586	100.0	22,265	100.0
Gender												
Male	194,889	85.3	52,186	76.3	48,201	96.7	17,146	88.8	62,661	91.4	14,695	66.0
Female	33,553	14.7	16,246	23.7	1,653	3.3	2,159	11.2	5,925	8.6	7,570	34.0
Race/Ethnic Group												
White	200,015	87.6	62,150	90.8	39,016	78.3	17,126	88.7	60,941	88.9	20,782	93.3
Asian or Pacific Islander	21,085	9.2	4,034	5.9	9,434	18.9	1,519	7.9	5,788	8.4	309	1.4
African American	3,077	1.3	1,158	1.7	535	1.1	241	1.2	579	0.8	563	2.5
Hispanic	3,584	1.6	927	1.4	728	1.5	359	1.9	1,109	1.6	461	2.1
American Indian	682	0.3	163	0.2	140	0.3	61	0.3	169	0.2	149	0.7
Underrepresented Minorities	7,343	3.2	2,248	3.3	1,403	2.9	661	3.4	1,857	2.6	1,173	5.3
Disability Status												
Disabled	1,484	0.6	289	0.4	264	0.5	199	1.0	517	0.8	216	1.0
Not Disabled	226,958	99.4	68,142	99.6	49,591	99.5	19,107	99.0	68,069	99.2	22,049	99.0

NOTE: This table is limited to individuals who were U.S. citizens in 1993, working in a field closely or somewhat related to their Ph.D. field. Disabled persons are defined as individuals who have severe difficulty seeing, hearing, walking, and/or lifting or are unable to perform these tasks. The underrepresented minorities include African Americans, Hispanics, and American Indians.

Source: 1993 Survey of Doctorate Recipients

TABLE A-1.2 Number of Employed Physical Science Doctorates (1960–89 graduates), by Field of Doctorate and Demographic Characteristics, 1993

Demographic Characteristics	Total #	Total %	Astronomy #	Astronomy %	Atmospheric Sci & Meteorology #	Atmospheric Sci & Meteorology %	Chemistry #	Chemistry %	Geological & Related Sci #	Geological & Related Sci %	Physics #	Physics %	Miscellaneous Physical Sci #	Miscellaneous Physical Sci %
Total	68,586	100.0	1,970	100.0	1,360	100.0	35,644	100.0	6,668	100.0	20,151	100.0	2,793	100.0
Gender														
Male	62,661	91.4	1,807	91.7	1,330	97.8	31,648	88.8	6,086	91.3	19,339	96.0	2,451	87.8
Female	5,925	8.6	163	8.3	30	2.2	3,996	11.2	582	8.7	812	4.0	342	12.2
Race/Ethnic Group														
White	60,941	88.9	1,812	92.0	1,243	91.4	31,143	87.4	6,301	94.5	17,780	88.2	2,662	95.3
Asian or Pacific Islander	5,788	8.4	136	6.9	97	7.1	3,312	9.3	244	3.7	1,890	9.4	109	3.9
African American	580	0.9	10	0.5	3	0.2	417	1.2	27	0.4	123	0.6	0	0.0
Hispanic	1,109	1.6	9	0.5	5	0.4	655	1.8	94	1.4	331	1.6	15	0.5
American Indian	169	0.2	3	0.2	12	0.9	117	0.3	2	0.0	27	0.1	8	0.3
Underrepresented Minorities	1,858	2.7	22	1.2	20	1.5	1,189	3.3	123	1.8	481	2.3	23	0.8
Disability														
Disabled	516	0.8	0	0.0	0	0.0	277	0.8	54	0.8	144	0.7	41	1.5
Not Disabled	68,069	99.2	1,970	100.0	1,360	100.0	35,366	99.2	6,614	99.2	20,007	99.3	2,752	98.5

NOTE: This table is limited to individuals who were U.S. citizens in 1993, working in a field closely or somewhat related to their Ph.D. field. Disabled persons are defined as individuals who have severe difficulty seeing, hearing, walking, and/or lifting or are unable to perform these tasks. The underrepresented minorities include African Americans, Hispanics, and American Indians.

Source: 1993 Survey of Doctorate Recipients

TABLE A-1.3 Number of Employed S&E Doctorates (1960–89 graduates), by Year of Doctorate and Demographic Characteristics, 1993

Demographic Characteristics	Total		1960–64		1965–69		1970–74		1975–79		1980–84		1985–89	
	#	%	#	%	#	%	#	%	#	%	#	%	#	%
Total	228,442	100.0	13,669	100.0	37,365	100.0	47,840	100.0	42,949	100.0	42,561	100.0	44,057	100.0
Gender														
Male	194,889	85.3	13,023	95.3	34,897	93.4	43,136	90.2	36,532	85.1	34,249	80.5	33,052	75.0
Female	33,553	14.7	646	4.7	2,469	6.6	4,704	9.8	6,417	14.9	8,313	19.5	11,005	25.0
Race/Ethnic Group														
White	200,015	87.6	12,790	93.6	33,502	89.7	41,649	87.1	36,530	85.1	36,672	86.2	38,872	88.2
Asian or Pacific Islander	21,085	9.2	757	5.5	2,971	8.0	4,579	9.6	4,796	11.2	4,476	10.5	3,505	8.0
African American	3,077	1.3	54	0.4	280	0.7	616	1.3	863	2.0	607	1.4	657	1.5
Hispanic	3,584	1.6	51	0.4	476	1.3	812	1.7	635	1.5	640	1.5	968	2.2
American Indian	682	0.3	16	0.1	136	0.4	184	0.4	126	0.3	166	0.4	55	0.1
Underrepresented Minorities	7,343	3.2	121	0.9	892	2.4	1,612	3.4	1,624	3.8	1,413	3.3	1,680	3.8
Disability														
Disabled	1,484	0.6	181	1.3	336	0.9	311	0.7	190	0.4	295	0.7	170	0.4
Not Disabled	226,958	99.4	13,488	98.7	37,029	99.1	47,529	99.3	42,759	99.6	42,267	99.3	43,887	99.6

NOTE: This table is limited to individuals who were U.S. citizens in 1993, working in a field closely or somewhat related to their Ph.D. field. Disabled persons are defined as individuals who have severe difficulty seeing, hearing, walking, and/or lifting or are unable to perform these tasks. The underrepresented minorities include African Americans, Hispanics, and American Indians.

Source: 1993 Survey of Doctorate Recipients

TABLE A-1.4 Number of Employed S&E Doctorates (1960–89 graduates), by Year of College Entrance

Demographic Characteristics	Total		Before 1950		1950–59		1960–69		1970–79		1980–89	
	#	%	#	%	#	%	#	%	#	%	#	%
Total	228,442	100.0	19,220	100.0	60,741	100.0	89,891	100.0	57,686	100.0	905	100.0
Gender												
Male	194,889	85.3	16,236	84.5	56,094	92.3	76,672	85.3	45,304	78.5	582	64.3
Female	33,553	14.7	2,984	15.5	4,646	7.6	13,219	14.7	12,382	21.5	323	35.7
Race/Ethnic Group												
White	200,015	87.6	16,682	86.8	53,119	87.5	78,112	86.9	51,292	88.9	811	89.6
Asian or Pacific Islander	21,085	9.2	1,737	9.0	6,066	10.0	8,681	9.7	4,523	7.8	77	8.5
African American	3,077	1.3	393	2.0	581	1.0	1,458	1.6	637	1.1	8	0.9
Hispanic	3,584	1.6	324	1.7	775	1.3	1,370	1.5	1,105	1.9	10	1.1
American Indian	682	0.3	84	0.4	199	0.3	270	0.3	128	0.2	0	0.0
Underrepresented Minorities	7,343	3.2	801	4.1	1,555	2.6	3,098	3.4	1,870	3.2	18	2.0
Disability Status												
Disabled	1,485	0.6	196	1.0	609	1.0	432	0.5	248	0.4	0	0.0
Not Disabled	226,958	99.4	19,025	99.0	60,132	99.0	89,459	99.5	57,438	99.6	905	100.0

NOTE: This table is limited to individuals who were U.S. citizens in 1993, working in a field closely or somewhat related to their Ph.D. field. Disabled persons are defined as individuals who have severe difficulty seeing, hearing, walking, and/or lifting or are unable to perform these tasks. The underrepresented minorities include African Americans, Hispanics, and American Indians.

Source: 1993 Survey of Doctorate Recipients.

TABLE A-1.5 Number of Underrepresented S&E Doctorates (1960–89 graduates), by Field of Doctorate, Sector of Employment, and Carnegie Classification, 1993

Sector of Employment	Total		Biological Sciences		Engineering		Mathematics & Computer Sciences		Physical Sciences		Psychology	
	#	%	#	%	#	%	#	%	#	%	#	%
Total	40,385	100.0	17,945	100.0	3,200	100.0	2,918	100.0	7,982	100.0	8,340	100.0
Sector												
Univ. & 4-yr. colleges	20,325	50.3	10,591	59.0	1,166	36.4	1,854	63.5	3,143	39.4	3,572	42.8
Other educational inst.	1,814	4.5	657	3.7	36	1.1	156	5.3	384	4.8	581	7.0
Private-for-profit	10,093	25.0	3,401	19.0	1,483	46.3	619	21.2	3,274	41.0	1,316	15.8
Self-employed	2,064	5.1	347	1.9	77	2.4	46	1.6	173	2.2	1,421	17.0
Private not-for-profit	1,988	4.9	1,005	5.6	74	2.3	80	2.7	248	3.1	581	7.0
Federal government	3,047	7.5	1,556	8.7	324	10.1	129	4.4	683	8.6	354	4.2
State and local govt.	1,029	2.5	363	2.0	40	1.3	34	1.2	78	1.0	515	6.2
Other	24	0.2	24	0.1	0	0.0	0	0.0	0	0.0	0	0.0
Carnegie Classification of Academic Institutions*												
Research universities	9,366	46.1	5,670	53.5	521	44.7	582	31.4	1,249	39.7	1,346	37.7
Doctorate-granting univ.	1,867	9.2	800	7.6	161	13.8	237	12.8	326	10.4	343	9.6
Comprehensive	4,105	20.2	1,473	13.9	250	21.4	611	33.0	767	24.4	1,005	28.1
Liberal arts	2,043	10.1	773	7.3	62	5.3	274	14.8	430	13.7	505	14.1
Medical & health schools	1,895	9.3	1,540	14.5	9	0.8	22	1.2	76	2.4	247	6.9
Other institutions	1,048	5.2	336	3.2	163	14.0	128	6.9	295	9.4	126	3.5

NOTE: This table is limited to individuals who were U.S. citizens in 1993, working in a field closely or somewhat related to their Ph.D. field. The underrepresented group is comprised of women, African Americans, Hispanics, American Indians, and disabled persons drawn from the categories in Table A-1.1, eliminating duplications.

*Includes individuals employed by universities and four-year colleges only.

SOURCE: 1993 Survey of Doctorate Recipients

TABLE A-1.6 Number of Underrepresented S&E Doctorates (1960–89 graduates) Employed in Academia, by Field of Doctorate and Academic Rank, 1993

Faculty Rank	Total #	Total %	Biological Sciences #	Biological Sciences %	Engineering #	Engineering %	Mathematics & Computer Sciences #	Mathematics & Computer Sciences %	Physical Sciences #	Physical Sciences %	Psychology #	Psychology %
Total	21,500	100.0	11,025	100.0	1,199	100.0	1,989	100.0	3,421	100.0	3,867	100.0
Full professor	5,392	25.1	2,171	19.7	338	28.2	772	38.8	979	28.6	1,133	29.3
Associate professor	5,629	26.2	2,708	24.6	343	28.6	681	34.2	840	24.6	1,057	27.3
Assistant professor	5,063	23.5	2,940	26.7	314	26.2	275	13.8	743	21.7	791	20.5
Instructor/lecturer	1,100	5.1	568	5.2	32	2.7	117	5.9	206	6.0	177	4.6
Adjunct faculty	657	3.1	354	3.2	56	4.7	43	2.2	62	1.8	142	3.7
Other	133	0.6	53	0.5	0	0.0	15	0.8	31	0.9	34	0.9
Not applicable	3,526	16.4	2,232	20.2	115	9.6	86	4.3	560	16.4	533	13.8

NOTE: This table is limited to individuals who were U.S. citizens in 1993, working in a field closely or somewhat related to their Ph.D. field. The underrepresented group is comprised of women, African Americans, Hispanics, American Indians, and disabled persons drawn from the categories in Table A-1.1, eliminating duplications.

SOURCE: 1993 Survey of Doctorate Recipients

TABLE A-1.7 Number of Underrepresented S&E Doctorates (1960–89 graduates), by Field of Doctorate and Supervision Responsibilities, 1993

Supervisory Duties	Total		Biological Sciences		Engineering		Mathematics & Computer Sciences		Physical Sciences		Psychology	
	#	%	#	%	#	%	#	%	#	%	#	%
Total	40,385	100.0	17,945	100.0	3,200	100.0	2,918	100.0	7,982	100.0	8,340	100.0
Supervises others	25,323	62.7	12,718	70.9	2,002	62.6	1,132	38.8	4,708	59.0	4,763	57.1
Does not supervise	15,062	37.3	5,228	29.1	1,198	37.4	1,785	61.2	3,274	41.0	3,577	42.9

NOTE: This table is limited to individuals who were U.S. citizens in 1993, working in a field closely or somewhat related to their Ph.D. field. The underrepresented group is comprised of women, African Americans, Hispanics, American Indians, and disabled persons drawn from categories in Table A-1.1, eliminating duplications.

SOURCE: 1993 Survey of Doctorate Recipients

TABLE A-1.8 Number of Underrepresented S&E Doctorates (1960–89 graduates), by Field of Doctorate and R&D Status, 1993

	Total		Biological Sciences		Engineering		Mathematics & Computer Sciences		Physical Sciences		Psychology	
Engaged in R&D?	#	%	#	%	#	%	#	%	#	%	#	%
Total	40,385	100.0	17,945	100.0	3,200	100.0	2,918	100.0	7,982	100.0	8,340	100.0
Conducting R&D	31,997	79.2	14,685	81.8	2,809	87.8	2,264	77.6	6,527	81.8	5,712	68.5
Not conducting R&D	8,387	20.8	3,260	18.2	390	12.2	654	22.4	1,455	18.2	2,628	31.5

NOTE: This table is limited to individuals who were U.S. citizens in 1993, working in a field closely or somewhat related to their Ph.D. field. The underrepresented group is comprised of women, African Americans, Hispanics, American Indians, and disabled persons drawn from the categories in Table A-1.1, eliminating duplications. Individuals conducting R&D spend 10 percent or more of their time engaged in research or development in a typical week.

SOURCE: 1993 Survey of Doctorate Recipients

TABLE A-2.1 Number of Employed Engineers with Master's in Engineering (1960–89 graduates), by Field of Degree and Demographic Characteristics, 1993

Demographic Characteristics	Total		Electrical & Computer Sys		Mechanical & Industrial		Civil & Architectural		Chemical		Other	
	#	%	#	%	#	%	#	%	#	%	#	%
Total	288,281	100.0	90,068	100.0	81,766	100.0	55,074	100.0	23,746	100.0	37,627	100.0
Gender												
Male	270,416	93.8	86,201	95.7	77,868	95.2	50,575	91.8	21,088	88.8	34,684	92.2
Female	17,865	6.2	3,867	4.3	3,898	4.8	4,499	8.2	2,658	11.2	2,943	7.8
Race/Ethnic Group												
White	245,638	85.2	75,846	84.2	69,378	84.8	45,909	83.4	19,810	83.4	34,696	92.2
Asian or Pacific Islander	36,465	12.6	11,648	12.9	10,574	12.9	8,514	15.5	3,259	13.7	2,472	6.6
African American	4,071	1.4	1,751	1.9	1,222	1.5	483	0.9	333	1.4	281	0.7
Hispanic	1,063	0.4	561	0.6	224	0.3	169	0.3	109	0.5	0	0.0
American Indian	532	0.2	146	0.2	151	0.2	0	0.0	235	1.0	0	0.0
Unknown	512	0.2	116	0.1	217	0.3	0	0.0	0	0.0	179	0.5
Underrepresented Minorities	5,666	2.0	2,458	2.7	1,597	2.0	652	1.2	677	2.9	281	0.7
Disability Status												
Disabled	1,292	0.4	188	0.2	293	0.4	474	0.9	54	0.2	283	0.8
Not Disabled	286,989	99.6	89,880	99.8	81,472	99.6	54,600	99.1	23,692	99.8	37,344	99.2

NOTE: This table is limited to individuals who were U.S. citizens in 1993, working in a field closely or somewhat related to their Ph.D. field. Disabled persons are defined as individuals who have severe difficulty seeing, hearing, walking, and/or lifting or are unable to perform these tasks. The underrepresented minorities include African Americans, Hispanics, and American Indians.

Source: 1993 National Survey of College Graduates

TABLE A-2.2 Number of Employed Engineers with Master's in Engineering (1960–89 graduates), by Year of Graduation and Demographic Characteristics, 1993

Demographic Characteristics	Total		1960–64		1965–69		1970–74		1975–79		1980–84		1985–89	
	#	%	#	%	#	%	#	%	#	%	#	%	#	%
Total	288,281	100.0	17,756	100.0	36,230	100.0	49,997	100.0	54,742	100.0	60,693	100.0	68,863	100.0
Gender														
Male	270,416	93.8	17,389	97.9	36,009	99.4	49,178	98.4	53,024	96.9	54,639	90.0	60,177	87.4
Female	17,865	6.2	366	2.1	221	0.6	819	1.6	1,718	3.1	6,054	10.0	8,686	12.6
Race/Ethnic Group														
White	245,638	85.2	15,573	87.7	31,893	88.0	42,569	85.1	45,705	83.5	49,549	81.6	60,348	87.6
Asian or Pacific Islander	36,465	12.6	1,969	11.1	4,169	11.5	6,848	13.7	7,951	14.5	9,355	15.4	6,173	9.0
African American	4,071	1.4	162	0.9	0	0.0	456	0.9	854	1.6	1,319	2.2	1,280	1.9
Hispanic	1,063	0.4	0	0.0	110	0.3	67	0.1	232	0.4	312	0.5	342	0.5
American Indian	532	0.2	52	0.3	0	0.0	0	0.0	0	0.0	99	0.2	381	0.6
Unknown	512	0.2	0	0.0	57	0.2	57	0.1	0	0.0	59	0.1	339	0.5
Underrepresented Minorities	5,666	2.0	214	1.2	110	0.3	523	1.0	1,086	2.0	1,730	2.9	2,003	3.0
Disability														
Disabled	1,292	0.4	254	1.4	121	0.3	135	0.3	285	0.5	187	0.3	310	0.5
Not Disabled	286,989	99.6	17,502	98.6	36,109	99.7	49,863	99.7	54,457	99.5	60,506	99.7	68,553	99.5

NOTE: This table is limited to individuals who were U.S. citizens in 1993, working in a field closely or somewhat related to their Ph.D. field. Disabled persons are defined as individuals who have severe difficulty seeing, hearing, walking, and/or lifting or are unable to perform these tasks. The underrepresented minorities include African Americans, Hispanics, and American Indians.

Source: 1993 National Survey of College Graduates

TABLE A-2.3 Number of Underrepresented Engineers with Master's in Engineering (1960–89 graduates), by Field of Degree and Sector of Employment, 1993

Sector of Employment	Total		Electrical & Computer Sys		Mechanical & Industrial		Civil & Architectural		Chemical		Other	
	#	%	#	%	#	%	#	%	#	%	#	%
Total	23,729	100.0	6,200	100.0	5,498	100.0	5,248	100.0	3,275	100.0	3,508	100.0
Sector												
Univ. & 4-yr. colleges	986	4.2	217	3.5	57	1.0	52	1.0	403	12.3	258	7.4
Other educational inst.	249	1.0	55	0.9	0	0.0	135	2.6	0	0.0	59	1.7
Private-for-profit	17,099	72.1	4,822	77.8	4,365	79.4	2,896	55.2	2,396	73.2	2,621	74.7
Self-employed	586	2.5	113	1.8	135	2.5	189	3.6	0	0.0	149	4.2
Private not-for-profit	265	1.1	118	1.9	94	1.7	0	0.0	0	0.0	53	1.5
Federal government	2,953	12.4	666	10.7	680	12.4	1,072	20.4	302	9.2	233	6.6
State and local govt.	1,278	5.4	159	2.6	52	0.9	756	14.4	174	5.3	136	3.9
Other	314	1.3	51	0.8	114	2.1	149	2.8	0	0.0	0	0.0

NOTE: This table is limited to individuals who were U.S. citizens in 1993, working in a field closely or somewhat related to their Ph.D. field. The underrepresented group is comprised of women, African Americans, Hispanics, American Indians, and disabled persons drawn from the categories in Table A-2.1, eliminating duplications.

SOURCE: 1993 National Survey of College Graduates

TABLE A-2.4 Number of Underrepresented Engineers with Master's in Engineering (1960–89 graduates), by Field of Degree and Professional Experience, 1993

Years of Professional Experience	Total		Electrical & Computer Sys		Mechanical & Industrial		Civil & Architectural		Chemical		Other	
	#	%	#	%	#	%	#	%	#	%	#	%
Total	23,729	100.0	6,200	100.0	5,498	100.0	5,248	100.0	3,275	100.0	3,508	100.0
1 to 4 years	1,782	7.5	282	4.5	757	13.8	233	4.4	510	15.6	0	0.0
5 to 10 years	7,477	31.5	1,605	25.9	2,171	39.5	1,736	33.1	995	30.4	969	27.6
Over 10 years	14,470	61.0	4,313	69.6	2,571	46.8	3,279	62.5	1,769	54.0	2,538	72.3

NOTE: This table is limited to individuals who were U.S. citizens in 1993, working in a field closely or somewhat related to their Ph.D. field. The underrepresented group is comprised of women, African Americans, Hispanics, American Indians, and disabled persons drawn from the categories in Table A-2.1, eliminating duplications.

SOURCE: 1993 National Survey of College Graduates

TABLE A-2.5 Number of Underrepresented Engineers with Master's in Engineering (1960–89 graduates), by Field of Degree and Occupation, 1993

Occupation	Total		Electrical & Computer Sys		Mechanical & Industrial		Civil & Architectural		Chemical		Other	
	#	%	#	%	#	%	#	%	#	%	#	%
Total	23,729	100.0	6,200	100.0	5,498	100.0	5,248	100.0	3,275	100.0	3,508	100.0
Manager*	2,700	11.4	603	9.7	276	5.0	1,119	21.3	240	7.3	462	13.2
Not a manager	21,029	88.6	5,597	90.3	5,222	95.0	4,130	78.7	3,035	92.7	3,045	86.8

NOTE: This table is limited to individuals who were U.S. citizens in 1993, working in a field closely or somewhat related to their Ph.D. field. The underrepresented group is comprised of women, African Americans, Hispanics, American Indians, and disabled persons drawn from the categories in Table A-2.1, eliminating duplications.

*This category includes top and mid-level managers, executives, and administrators (people who manage other managers).

SOURCE: 1993 National Survey of College Graduates

TABLE A-2.6 Number of Underrepresented Engineers with Master's in Engineering (1960–89 graduates), by Field of Degree and R&D Status, 1993

Research & Development Status	Total		Electrical & Computer Sys		Mechanical & Industrial		Civil & Architectural		Chemical		Other	
	#	%	#	%	#	%	#	%	#	%	#	%
Total	23,729	100.0	6,200	100.0	5,498	100.0	5,248	100.0	3,275	100.0	3,508	100.0
Conducting R&D	14,029	59.1	3,535	57.0	3,480	63.3	2,398	45.7	2,694	82.3	1,922	54.8
Not conducting R&D	9,700	40.9	2,665	43.0	2,018	36.7	2,850	54.3	581	17.7	1,586	45.2

NOTE: This table is limited to individuals who were U.S. citizens in 1993, working in a field closely or somewhat related to their Ph.D. field. The underrepresented group is comprised of women, African Americans, Hispanics, American Indians, and disabled persons drawn from the categories in Table A-2.1, eliminating duplications.

SOURCE: 1993 National Survey of College Graduates

TABLE A-3.1 Number of Recent Ph.D. Scientists and Engineers (1990–95 graduates), by Field of Doctorate and Demographic Characteristics

Demographic Characteristics	Total #	Total %	Biological Sciences #	Biological Sciences %	Engineering #	Engineering %	Mathematics & Computer Sciences #	Mathematics & Computer Sciences %	Physical Sciences #	Physical Sciences %	Psychology #	Psychology %
Total	55,075	100.0	18,870	100.0	11,119	100.0	4,797	100.0	14,844	100.0	5,445	100.0
Gender												
Male	38,415	69.8	11,111	58.9	9,501	85.4	3,734	77.8	11,628	78.3	2,441	44.8
Female	16,660	30.2	7,759	41.1	1,618	14.6	1,063	22.2	3,216	21.7	3,004	55.2
Race/Ethnic Group												
White	51,281	93.1	17,534	92.9	10,312	92.7	4,558	95.0	13,946	94.0	4,931	90.6
Asian or Pacific Islander	1,378	2.5	502	2.7	348	3.1	104	2.2	343	2.3	81	1.5
African American	920	1.7	304	1.6	193	1.7	52	1.1	157	1.1	214	3.9
Hispanic	1,301	2.4	465	2.5	231	2.1	71	1.5	343	2.3	191	3.5
American Indian	195	0.4	65	0.3	35	0.3	12	0.3	55	0.4	28	0.5
Underrepresented Minorities	2,416	4.5	834	4.4	459	4.1	135	2.9	555	3.8	433	7.9
Disability Status												
Disabled	664	1.2	218	1.2	119	1.1	64	1.3	186	1.3	77	1.4
Not Disabled	54,411	98.8	18,652	98.8	11,000	98.9	4,733	98.7	14,658	98.7	5,368	98.6

NOTE: This table is limited to individuals who were U.S. citizens at the time of the Ph.D. award. Disabled persons are defined as those individuals who indicated that they had a disability. The underrepresented minorities include African Americans, Hispanics, and American Indians.

Source: Survey of Earned Doctorates

TABLE A-3.2 Number of Recent Physical Science Doctorates (1990–95 graduates), by Field of Doctorate and Demographic Characteristics

Demographic Characteristics	Total		Astronomy		Atmospheric Sci & Meteorology		Chemistry		Geological & Related Sci		Physics		Miscellaneous Physical Sci	
	#	%	#	%	#	%	#	%	#	%	#	%	#	%
Total	14,844	100.0	573	100.0	312	100.0	7,374	100.0	1,880	100.0	3,885	100.0	820	100.0
Gender														
Male	11,628	78.3	483	84.3	266	85.3	5,376	72.9	1,428	76.0	3,500	90.1	575	70.1
Female	3,216	21.7	90	15.7	46	14.7	1,998	27.1	452	24.0	385	9.9	245	29.9
Race/Ethnic Group														
White	13,946	94.0	551	96.2	306	98.1	6,831	92.6	1,817	96.6	3,655	94.1	786	95.9
Asian or Pacific Islander	343	2.3	8	1.4	0	0.0	206	2.8	14	0.7	104	2.7	11	1.3
African American	157	1.1	4	0.7	3	1.0	101	1.4	10	0.5	33	0.8	6	0.7
Hispanic	343	2.3	9	1.6	3	1.0	205	2.8	32	1.7	80	2.1	14	1.7
American Indian	55	0.3	1	0.2	0	0.0	31	0.4	7	0.4	13	0.3	3	0.4
Underrepresented Minorities	555	3.7	14	2.5	6	2.0	337	4.6	49	2.6	126	3.2	23	2.8
Disability														
Disabled	186	1.3	6	1.0	3	1.0	92	1.2	23	1.2	46	1.2	16	2.0
Not Disabled	14,658	98.7	567	99.0	309	99.0	7,282	98.8	1,857	98.8	3,839	98.8	804	98.0

NOTE: This table is limited to individuals who were U.S. citizens at the time of the Ph.D. award. Disabled persons are defined as those individuals who indicated that they had a disability. The underrepresented minorities include African Americans, Hispanics, and American Indians.

Source: Survey of Earned Doctorates

TABLE A-3.3 Number of Underrepresented Recent S&E Doctorates (1990–95 graduates), by Field of Doctorate and Geographic Location of Ph.D. Institution

Geographic Location of Institution	Total		Biological Sciences		Engineering		Mathematics & Computer Sciences		Physical Sciences		Psychology	
	#	%	#	%	#	%	#	%	#	%	#	%
Total	18,584	100.0	8,337	100.0	2,075	100.0	1,204	100.0	3,765	100.0	3,203	100.0
Location												
New England	1,780	9.6	868	10.4	183	8.8	88	7.3	393	10.4	248	7.7
Middle Atlantic	3,004	16.2	1,381	16.6	336	16.2	209	17.4	506	13.4	572	17.9
East North Central	3,079	16.6	1,378	16.5	374	18.0	173	14.4	637	16.9	517	16.1
West North Central	1,105	5.9	557	6.7	95	4.6	61	5.1	197	5.2	195	6.1
South Atlantic	3,019	16.2	1,399	16.8	343	16.5	235	19.5	555	14.7	487	15.2
East South Central	674	3.6	358	4.3	65	3.1	50	4.2	78	2.1	123	3.8
West South Central	1,447	7.8	629	7.5	168	8.1	117	9.7	296	7.9	237	7.4
Mountain	1,080	5.8	400	4.8	108	5.2	79	6.6	336	8.9	157	4.9
Pacific	3,396	18.3	1,367	16.4	403	19.4	192	15.9	767	20.4	667	20.8

NOTE: This table is limited to individuals who were U.S. citizens at the time of the Ph.D. award. The underrepresented group is comprised of women, African Americans, Hispanics, American Indians, and disabled persons drawn from the categories in Table A-3.1, eliminating duplications.

Source: Survey of Earned Doctorates

TABLE A-3.4 Number of Underrepresented Recent S&E Doctorates (1990–95 graduates), by Field of Doctorate and Carnegie Classification

Carnegie Classification for Doctorate Institution	Total		Biological Sciences		Engineering		Mathematics & Computer Sciences		Physical Sciences		Psychology	
	#	%	#	%	#	%	#	%	#	%	#	%
Total	18,584	100.0	8,337	100.0	2,075	100.0	1,204	100.0	3,765	100.0	3,203	100.0
Research Univ. I	13,427	72.3	6,318	75.8	1,696	81.7	817	67.9	2,918	77.5	1,678	52.4
Research Univ. II	1,908	10.3	675	8.1	199	9.6	157	13.0	434	11.5	443	13.8
Doctoral Univ. I	1,386	7.5	316	3.8	75	3.6	136	11.3	179	4.8	680	21.2
Doctoral Univ. II	800	4.3	297	3.6	72	3.5	83	6.9	184	4.9	164	5.1
Medical Schools & Centers	684	3.7	650	7.8	3	0.1	1	0.1	9	0.2	21	0.7
Other Institutions	379	2.0	81	1.0	30	1.4	10	0.8	41	1.1	217	6.8

NOTE: This table is limited to individuals who were U.S. citizens at the time of the Ph.D. award. The underrepresented group is comprised of women, African Americans, Hispanics, American Indians, and disabled persons drawn from the categories in Table A-3.1, eliminating duplications.

Source: Survey of Earned Doctorates

TABLE A-3.5 Number of Underrepresented Recent S&E Doctorates (1990–95 graduates), by Field of Doctorate and Locus of Institutional Control

Locus of Control	Total		Biological Sciences		Engineering		Mathematics & Computer Sciences		Physical Sciences		Psychology	
	#	%	#	%	#	%	#	%	#	%	#	%
Total	18,584	100.0	8,337	100.0	2,075	100.0	1,204	100.0	3,765	100.0	3,203	100.0
Publicly controlled	12,468	67.1	5,637	67.6	1,312	63.2	829	68.9	2,667	70.8	2,023	63.2
Privately controlled	6,116	32.9	2,700	32.4	763	36.8	375	31.1	1,098	29.2	1,180	36.8

NOTE: This table is limited to individuals who were U.S. citizens at the time of the Ph.D. award. The underrepresented group is comprised of women, African Americans, Hispanics, American Indians, and disabled persons drawn from the categories in Table A-3.1, eliminating duplications.

Source: Survey of Earned Doctorates

TABLE A-4.1 Number of Engineers with Recent Master's in Engineering (1990–93 graduates), by Field of Degree and Demographic Characteristics, 1993

Demographic Characteristics	Total		Electrical & Computer Sys		Mechanical & Industrial		Civil & Architectural		Chemical		Other	
	#	%	#	%	#	%	#	%	#	%	#	%
Total	55,766	100.0	21,510	100.0	18,530	100.0	6,815	100.0	3,404	100.0	5,508	100.0
Gender												
Male	48,396	86.8	18,460	85.8	16,445	88.7	5,776	84.8	3,016	88.6	4,698	85.3
Female	7,370	13.2	3,050	14.2	2,084	11.2	1,039	15.2	388	11.4	809	14.7
Race/Ethnic Group												
White	48,499	87.0	18,163	84.4	16,371	88.3	6,411	94.1	2,898	85.1	4,655	84.5
Asian or Pacific Islander	4,509	8.1	2,523	11.7	1,159	6.3	165	2.4	274	8.0	387	7.0
African American	2,063	3.7	663	3.1	686	3.7	182	2.7	180	5.3	352	6.4
Hispanic	219	0.4	0	0.0	107	0.6	57	0.8	0	0.0	55	1.0
American Indian	104	0.2	0	0.0	53	0.3	0	0.0	51	1.5	0	0.0
Unknown	372	0.6	161	0.7	152	0.8	0	0.0	0	0.0	59	1.1
Underrepresented Minorities	2,386	4.3	663	3.1	846	4.6	239	3.5	231	6.8	407	7.4
Disability Status												
Disabled	0	0.0	0	0.0	0	0.0	0	0.0	0	0.0	0	0.0
Not Disabled	55,766	100.0	21,510	100.0	18,530	100.0	6,815	100.0	3,404	100.0	5,508	100.0

NOTE: This table is limited to individuals who were U.S. citizens in 1993, working in a field closely or somewhat related to their field of study. Disabled persons are defined as individuals who have severe difficulty seeing, hearing, walking, and/or lifting or are unable to perform these tasks. The underrepresented minorities include African Americans, Hispanics, and American Indians.

Source: 1993 National Survey of College Graduates

TABLE A-4.2 Number of Underrepresented Engineers with Recent Master's in Engineering (1990–93 graduates), by Field of Degree and Geographic Location, 1993

Geographic Location of Institution	Total		Electrical & Computer Sys		Mechanical & Industrial		Civil & Architectural		Chemical		Other	
	#	%	#	%	#	%	#	%	#	%	#	%
Total	9,346	100.0	3,464	100.0	2,821	100.0	1,278	100.0	619	100.0	1,164	100.0
Location												
New England	782	8.4	551	15.9	55	1.9	53	4.1	51	8.2	72	6.2
Middle Atlantic	2,510	26.9	1,019	29.4	882	31.3	374	29.3	115	18.6	120	10.3
East North Central	1,019	10.9	534	15.4	292	10.4	130	10.2	0	0.0	63	5.4
West North Central	271	2.9	142	4.1	73	2.6	56	4.4	0	0.0	0	0.0
South Atlantic	2,022	21.6	347	10.0	886	31.4	321	25.1	73	11.8	395	33.9
East South Central	207	2.2	0	0.0	207	7.3	0	0.0	0	0.0	0	0.0
West South Central	125	1.3	72	2.1	53	1.9	0	0.0	0	0.0	0	0.0
Mountain	666	7.1	145	4.2	73	2.6	0	0.0	133	21.5	315	27.1
Pacific	1,742	18.6	653	18.9	300	10.6	343	26.8	247	39.9	199	17.1

NOTE: This table is limited to individuals who were U.S. citizens in 1993, working in a field closely or somewhat related to their field of study. The underrepresented group is comprised of women, African Americans, Hispanics, American Indians, and disabled persons drawn from the categories in Table A-4.1, eliminating duplications.

Source: 1993 National Survey of College Graduates

TABLE A-5.1 ONR S&E Work Force by Field of Employment, Race/Ethnicity, and Gender

Race/Gender	Total		Biological Sciences		Engineering		Mathematics & Computer Sciences		Physical Sciences		Cognitive Sciences	
	#	%	#	%	#	%	#	%	#	%	#	%
Total	150	100.0	10	100.0	50	100.0	15	100.0	67	100.0	8	100.0
Male	111	74.0	5	50.0	41	82.0	7	46.7	54	80.6	4	50.0
Female	16	10.7	3	30.0	5	10.0	5	33.3	2	3.0	1	12.5
SrExec	23	15.3	2	20.0	4	16.4	3	20.0	11	16.4	3	37.5
White	137	100.0	10	100.0	40	100.0	14	100.0	65	100.0	8	100.0
Male	101	73.7	5	50.0	33	82.5	7	50.0	52	80.0	4	50.0
Female	13	9.5	3	30.0	3	7.5	4	28.6	2	3.1	1	12.5
SrExec	23	16.8	2	20.0	4	10.0	3	21.4	11	16.9	3	37.5
Asian American/ African American	13	100.0	0	0.0	10	100.0	1	100.0	2	100.0	0	0.0
Male	10	76.9	0	0.0	8	80.0	0	0.0	2	100.0	0	0.0
Female	3	23.1	0	0.0	2	20.0	1	100.0	0	0.0	0	0.0

NOTE: Field of employment is derived from the federal codes established for science and engineering occupations by the Office of Personnel Management, as follows: Biological Sciences (series 04xx), Engineering (Series 08xx), Mathematics and Computer Sciences (Series 15xx), Physical Sciences (Series 13xx), and Cognitive Sciences (Series 0180).

Source: Human Resources Office, Office of Naval Research, 1996

TABLE A-5.2 ONR S&E Work Force by Year of College Entrance

Category	Total		Before 1950		1950–59		1960–69		1970–79		1980–89		Unknown	
	#	%	#	%	#	%	#	%	#	%	#	%	#	%
Total	150	100.0	13	100.0	41	100.0	58	100.0	15	100.0	7	100.0	16	100.0
Gender														
Male	134	89.3	12	92.3	38	92.7	53	91.4	12	80.0	3	42.9	16	100.0
Female	16	10.7	1	7.7	3	7.3	5	8.6	3	20.0	4	57.1	0	0.0
Grade														
GS13	8	5.3	1	7.7	1	2.4	2	3.4	1	6.7	3	42.9	0	0.0
GS14	28	18.7	2	15.4	2	4.9	9	15.5	7	46.7	3	42.9	5	31.3
GS15	91	60.7	8	61.5	29	70.7	38	65.5	6	40.0	1	14.3	9	56.3
SrExec	23	15.3	2	15.4	9	22.0	9	15.5	1	6.7	0	0.0	2	12.5

Source: Human Resources Office, Office of Naval Research, 1996

TABLE A-5.3 ONR S&E Work Force by Highest Degree Earned, Grade, and Gender

Grade/Gender	Total		< Bachelor's		Bachelor's		Master's		Doctorate	
	#	%	#	%	#	%	#	%	#	%
Total	140	100.0	6	100.0	9	100.0	31	100.0	94	100.0
Male	125	89.3	6	100.0	7	77.8	26	83.9	86	91.5
Female	15	10.7	0	0.0	2	22.2	5	16.1	8	8.5
GS13	7	100.0	0	0.0	0	0.0	4	100.0	3	100.0
Male	3	42.9	0	0.0	0	0.0	1	25.0	2	66.7
Female	4	57.1	0	0.0	0	0.0	3	75.0	1	33.3
GS14	27	100.0	1	100.0	3	100.0	10	100.0	13	100.0
Male	24	88.9	1	100.0	2	66.7	9	90.0	12	92.3
Female	3	11.1	0	0.0	1	33.3	1	10.0	1	7.7
GS15	84	100.0	5	100.0	6	100.0	16	100.0	57	100.0
Male	76	90.5	5	100.0	5	83.3	15	93.8	51	89.5
Female	8	9.5	0	0.0	1	16.7	1	6.3	6	10.5
SrExec	22	100.0	0	0.0	0	0.0	1	100.0	21	100.0

NOTE: There are 10 records with missing values for highest degree earned.

Source: Human Resources Office, Office of Naval Research, 1996

TABLE A-5.4 ONR S&E Work Force by Field of Employment, Grade, and Gender

Grade/Gender	Total		Biological Sciences		Engineering		Mathematics & Computer Sciences		Physical Sciences		Cognitive Sciences	
	#	%	#	%	#	%	#	%	#	%	#	%
Total	150	100.0	10	100.0	50	100.0	15	100.0	67	100.0	8	100.0
Male	134	89.3	7	70.0	45	90.0	10	66.7	65	97.0	7	87.5
Female	16	10.7	3	30.0	5	10.0	5	33.3	2	3.0	1	12.5
GS13	8	100.0	1	100.0	4	100.0	2	100.0	1	100.0	0	0.0
Male	3	37.5	1	100.0	1	25.0	0	0.0	1	100.0	0	0.0
Female	5	62.5	0	0.0	3	75.0	2	100.0	0	0.0	0	0.0
GS14	28	100.0	0	0.0	8	100.0	2	100.0	16	100.0	2	100.0
Male	25	89.3	0	0.0	6	75.0	2	100.0	15	93.8	2	100.0
Female	3	10.7	0	0.0	2	25.0	0	0.0	1	6.3	0	0.0
GS15	91	100.0	7	100.0	34	100.0	8	100.0	39	100.0	3	100.0
Male	83	91.2	4	57.1	34	100.0	5	62.5	38	97.4	2	66.7
Female	8	8.8	3	42.9	0	0.0	3	37.5	1	2.6	1	33.3
SrExec	23	100.0	2	100.0	4	100.0	3	100.0	11	100.0	3	100.0

NOTE: Field of employment is derived from the federal codes established for science and engineering occupations by the Office of Personnel Management, as follows: Biological Sciences (series 04xx), Engineering (Series 08xx), Mathematics and Computer Sciences (Series 15xx), Physical Sciences (Series 13xx), and Cognitive Sciences (Series 0180).

Source: Human Resources Office, Office of Naval Research, 1996

TABLE A-5.5 ONR Start Date by Grade and Gender

Grade/Gender	Total		1965–79		1980–84		1985–89		1990–95	
	#	%	#	%	#	%	#	%	#	%
Total	150	100.0	10	100.0	32	100.0	52	100.0	56	100.0
Male	134	89.3	10	100.0	30	93.8	51	98.1	43	76.8
Female	16	10.7	0	0.0	2	6.3	1	1.9	13	23.2
GS13	8	100.0	0	0.0	0	0.0	1	100.0	7	100.0
Male	3	37.5	0	0.0	0	0.0	1	100.0	2	28.6
Female	5	62.5	0	0.0	0	0.0	0	0.0	5	71.4
GS14	28	100.0	2	100.0	1	100.0	8	100.0	17	100.0
Male	25	89.3	2	100.0	1	100.0	8	100.0	14	82.4
Female	3	10.7	0	0.0	0	0.0	0	0.0	3	17.6
GS15	91	100.0	5	100.0	21	100.0	40	100.0	25	100.0
Male	83	91.2	5	100.0	19	90.5	39	97.5	20	80.0
Female	8	8.8	0	0.0	2	9.5	1	2.5	5	20.0
SrExec	23	100.0	3	100.0	10	100.0	3	100.0	7	100.0

Source: Human Resources Office, Office of Naval Research, 1996

TABLE A-5.6 Target Grade for GS13 and GS14 ONR S&E Employees by Gender

Gender	Total		Target GS13		Target GS14		Target GS15	
	#	%	#	%	#	%	#	%
Total	36	100.0	3	100.0	15	100.0	18	100.0
Male	28	77.8	1	33.3	11	73.3	16	88.9
Female	8	22.2	2	66.7	4	26.7	2	11.1

Source: Human Resources Office, Office of Naval Research, 1996

TABLE A-5.7 Median Salaries of the ONR S&E Program Officers by Grade and Gender

ONR Grade/Number	Male Program Officers	Female Program Officers
Total		
Median Salary	90,640	83,475
Number of Individuals	111	16
GS13		
Median Salary	59,917	59,917
Number of Individuals	3	5
GS14		
Median Salary	76,012	74,969
Number of Individuals	25	3
GS15		
Median Salary	94,053	94,104
Number of Individuals	83	8

Source: Human Resources Office, Office of Naval Research, 1996

TABLE A-5.8 ONR S&E Work Force by the Most Recent and Second Most Recent Previous Employer

Previous Employer	Total		Male Program Officers		Female Program Officers		SrExec	
	#	%	#	%	#	%	#	%
Total	150	100.0	111	100.0	16	100.0	23	100.0
Most recent previous employer								
Academia	15	10.0	12	10.8	1	6.3	2	8.7
Industry	17	11.3	11	9.9	2	12.5	4	17.4
Navy	91	60.7	67	60.4	8	50.0	16	69.6
Other federal	27	18.0	21	18,9	5	31.3	1	4.3
Second most recent previous employer								
Academia	27	18.0	20	18.0	3	18.8	4	17.4
Industry	20	13.3	14	12.6	3	18.8	3	13.0
Navy	59	39.3	46	41.4	3	18.8	10	43.5
Other federal	44	29.3	31	27.9	7	43.8	6	26.1

Source: Human Resources Office, Office of Naval Research, 1996

TABLE A-5.9 ONR Employee Separations (1988–95) by Year, Gender, Race, and Grade

Category	Total	1988	1989	1990	1991	1992	1993	1994	1995
Total	87	14	7	9	9	11	7	20	10
Gender									
Male	82	14	6	8	9	10	7	19	9
Female	5	0	1	1	0	1	0	1	1
Race									
White	83	13	7	8	8	11	7	20	9
Asian American	2	0	0	0	1	0	0	0	1
African American	2	1	0	1	0	0	0	0	0
Grade									
GS13	2	0	0	0	0	1	0	1	0
GS14	20	4	0	4	5	1	2	3	1
GS15	46	6	5	4	3	7	3	10	8
SrExec	19	4	2	1	1	2	2	6	1

Source: Human Resources Office, Office of Naval Research, 1996

TABLE A-5.10 ONR Most Recent Hires (1994–95) by Gender, Field of Employment, Grade at Hire, Most Recent and Second Most Recent Previous Employer

Characteristics	Total		Male		Female	
	#	%	#	%	#	%
Total	18	100.0	11	100.0	7	100.0
Field of employment						
Biological Sciences	1	5.6	1	9.1	0	0.0
Engineering	6	33.3	4	36.4	2	28.6
Mathematics and Computer Sciences	6	33.3	3	27.3	3	42.9
Physical Sciences	5	27.8	3	27.3	2	28.6
Grade at hire						
GS13	6	33.3	2	18.2	4	57.1
GS14	6	33.3	5	45.5	1	14.3
GS15	5	27.8	3	27.3	2	28.6
SrExec	1	5.6	1	9.1	0	0.0
Most recent previous employer						
Industry	1	5.6	1	9.1	0	0.0
Navy	11	61.1	6	54.5	5	71.4
Other federal	6	33.3	4	36.4	2	28.6
Second most recent previous employer						
Academia	2	11.1	1	9.1	1	14.3
Industry	3	16.7	1	9.1	2	28.6
Navy	7	38.9	6	54.5	1	14.3
Other federal	6	33.3	3	27.3	3	42.9

NOTE: Field of employment is derived from the federal codes established for science and engineering occupations by the Office of Personnel Management, as follows: Biological Sciences (series 04xx), Engineering (Series 08xx), Mathematics and Computer Sciences (Series 15xx), and Physical Sciences (Series 13xx).

Source: Human Resources Office, Office of Naval Research, 1996

TABLE A-6.1 NRL S&E Employees by Field of Employment and Demographic Characteristics

Demographic Characteristics	Total		Biological Sciences		Engineering		Mathematics & Computer Sciences		Physical Sciences		Cognitive Sciences	
	#	%	#	%	#	%	#	%	#	%	#	%
Total	1,677	100.0	8	100.0	631	100.0	144	100.0	885	100.0	9	100.0
Gender												
Male	1,507	89.9	6	75.0	581	92.1	108	75.0	807	91.2	5	55.6
Female	170	10.1	2	25.0	50	7.9	36	25.0	78	8.8	4	44.4
Race/Ethnic Group												
White	1,491	88.9	7	87.5	540	85.6	133	92.4	802	90.6	9	100.0
Asian or Pacific Islander	142	8.5	1	12.5	71	11.3	5	3.5	65	7.3	0	0.0
African American	22	1.3	0	0.0	11	1.7	5	3.5	6	0.7	0	0.0
Hispanic	16	1.0	0	0.0	6	1.0	1	0.7	9	1.0	0	0.0
American Indian	6	0.3	0	0.0	3	0.5	0	0.0	3	0.3	0	0.0
Underrepresented Minorities	44	2.6	0	0.0	20	3.2	6	4.2	18	2.0	0	0.0
Disability Status												
Disabled	127	7.6	2	25.0	48	7.6	9	6.3	66	7.5	2	22.2
Not Disabled	1,550	92.4	6	75.0	583	92.4	135	93.8	819	92.5	7	77.8

NOTE: Field of employment is derived from the federal codes established for science and engineering occupations by the Office of Personnel Management, as follows: Biological Sciences (series 04xx), Engineering (Series 08xx), Mathematics and Computer Sciences (Series 15xx), Physical Sciences (Series 13xx), and Cognitive Sciences (Series 0180). Disabled persons are defined as those individuals who indicated that they had a disability.

Source: Naval Research Laboratory, 1996

TABLE A-6.2 NRL S&E Employees by Field of Employment, Grade, and Gender

Occupation, Status/Gender	Total		Biological Sciences		Engineering		Mathematics & Computer Sciences		Physical Sciences		Cognitive Sciences	
	#	%	#	%	#	%	#	%	#	%	#	%
Total	1,677	100.0	8	100.0	631	100.0	144	100.0	885	100.0	9	100.0
Gender												
Male	1,507	89.9	6	75.0	581	92.1	108	75.0	807	91.2	5	55.6
Female	170	10.1	2	25.0	50	7.9	36	25.0	78	8.8	4	44.4
GS12	477	100.0	2	100.0	227	100.0	48	100.0	166	100.0	4	100.0
Male	369	82.6	2	100.0	199	87.7	30	62.5	136	81.9	2	50.0
Female	78	17.4	0	0.0	28	12.3	18	37.5	30	18.1	2	50.0
GS13	512	100.0	4	100.0	185	100.0	55	100.0	265	100.0	3	100.0
Male	453	88.5	2	50.0	166	89.7	42	76.4	240	90.6	3	100.0
Female	59	11.5	2	50.0	19	10.3	13	23.6	25	9.4	0	0.0
GS14	387	100.0	2	100.0	127	100.0	29	100.0	227	100.0	2	100.0
Male	364	94.1	2	100.0	124	97.6	25	86.2	213	93.8	0	0.0
Female	23	5.9	0	0.0	3	2.4	4	13.8	14	6.2	2	100.0
GS15	287	100.0	0	0.0	86	100.0	12	100.0	189	100.0	0	0.0
Male	281	97.9	0	0.0	86	100.0	11	91.7	184	97.4	0	0.0
Female	6	2.1	0	0.0	0	0.0	1	8.3	5	2.6	0	0.0
SrExec	44	100.0	0	0.0	6	100.0	0	0.0	38	100.0	0	0.0
Male	40	90.9	0	0.0	6	100.0	0	0.0	34	89.5	0	0.0
Female	4	9.1	0	0.0	0	0.0	0	0.0	4	10.5	0	0.0

NOTE: Field of employment is derived from the federal codes established for science and engineering occupations by the Office of Personnel Management, as follows: Biological Sciences (series 04xx), Engineering (Series 08xx), Mathematics and Computer Sciences (Series 15xx), Physical Sciences (Series 13xx), and Cognitive Sciences (Series 0180).

Source: Naval Research Laboratory, 1996

TABLE A-6.3 NRL S&E Employees by Grade and Race

Grade/Race	Total		GS12		GS13		GS14		GS15		SrExec	
	#	%	#	%	#	%	#	%	#	%	#	%
Total	1,677	100.0	447	100.0	512	100.0	387	100.0	287	100.0	44	100.0
Race/Ethnicity												
White	1,491	88.9	367	82.1	469	91.6	342	88.4	272	94.8	41	93.2
Asian	142	8.5	58	13.0	34	6.6	37	9.6	11	3.8	2	4.5
African American	22	1.3	12	2.7	6	1.2	1	0.3	2	0.7	1	2.3
Hispanic	16	1.0	8	1.8	2	0.4	5	1.3	1	0.3	0	0.0
American Indian	6	0.4	2	0.4	1	0.2	2	0.5	1	0.3	0	0.0
Underrepresented Minorities	44	2.7	22	4.9	9	1.8	8	2.1	4	1.3	1	2.3

NOTE: The underrepresented minorities include African Americans, Hispanics, and American Indians.

Source: Naval Research Laboratory, 1996

APPENDIX B

RECENT ONR HIRING ACTIONS

To analyze ONR's recruitment and hiring practices, the committee examined data from 13 "case files" documenting the hires of scientists and engineers between January 1, 1993, and December 31, 1995.

The hiring actions were grouped into three types: **no recruitment**, **limited recruitment**, and **open recruitment**. No recruitment (2 hires) means that no vacancy announcement was prepared or advertised; the individual selected was the only candidate. Limited recruitment (4 hires) means that applications were limited to current Navy or DoD employees. Open recruitment (7 hires) means that the position was open to all qualified applicants, including those in industry and academia.

The limited and open recruitment practices applied to various geographic "areas of consideration": either the Washington, D.C. metropolitan area or nationwide. Thus a position which was open to qualified applicants in every sector of employment (academia, government, or industry) might have been advertised in the Washington, D.C. metropolitan area only.

There were nine case files where the job vacancy was advertised and the number of applicants known. Out of a total of 210 applicants for those nine positions, there were 106 qualified applicants. Data were not available on the race or disability status of the job applicants because those items are not included on the federal application form (SF171). ONR does ask applicants to fill out a separate (voluntary) form which gives this information, but none of the files examined had that form in them.

SUMMARY DATA ON QUALIFIED APPLICANTS*

Highest Level of Education Completed

	Total	B.S.	Master's	Ph.D.	Unknown
Total	106	29	38	38	1
Male	91	23	33	34	1
Female	14	6	5	3	
Unknown	1			1	

Employment at Time of Application

	Total	Government	Academia	Industry	Other	Unknown
Total	106	82	7	12	4	1
Male	91	69	6	11	4	1
Female	14	13		1		
Unknown	1		1			

Employment at Time of Application (Government only)

	Total	Nonmilitary	Military	Military-Navy	Military-ONR
Total	82	13	9	55	5
Male	69	12	9	44	4
Female	13	1		11	1

*Data is drawn from the nine case files where the job vacancy was advertised and the number of applications is known.

APPENDIX C

SUMMARY OF INTERVIEWS WITH ONR PROGRAM OFFICERS AND SENIOR EXECUTIVES

PART I: BACKGROUND AND METHODOLOGY

SCOPE

This report, prepared in consultation with Karen Bogart who conducted the interviews, summarizes the results of interviews with program officers and senior executives in the science and engineering work force in the headquarters of the Office of Naval Research in Arlington, Virginia. The first set of 45 interviews was conducted with a sample of women, minority males, and white males employed as GS13, GS14, and GS15 program officers. The second set of 26 interviews was conducted with senior executives and division and department heads, all of whom were white males.

THE STUDY SAMPLE

Program Officers

Study participants were drawn from the ONR science and engineering (S&E) work force, whose members numbered 150 in February 1996. The original study sample encompassed all women, all minority males* (Asian Americans and African Americans), and a sample of white males selected to match the ONR job codes (fields of specializations) from which minorities and women were drawn. Sample participants initially included 16 women, 9 minority males, and 26 white males for a total of 51. In March 1996, two participants who had inadvertently been miscoded were added, bringing the total study sample to 53.

Members of the study sample were contacted in November 1995 and asked to participate in a self-report survey conducted by mail. In the original mailing they were also advised that they would be contacted at a later date to schedule a 90-minute follow-up interview. In January 1996 prospective interview participants were contacted by e-mail and voice mail and asked to participate in an interview.

Senior Executives

In April 1996 all 27 senior executives at ONR were added to the study sample. This group was made up of the 19 members of the official Senior Executive Service (SES) and 8 senior managers. The SES members included 5 department directors who together constitute the Science and Technology Advisory Board (STAB). The department directors develop and implement science policy and represent ONR to Congress and to the Pentagon. Below the department directors were 14 SES members whose numbers included 12 division heads and 2 associates. The division heads have primary responsibility for recruiting, hiring, and promoting program officers. They administer people rather than programs, in contrast to the program officers whom they supervise. Among the senior managers who participated in the interviews were 4 chief scientists (GS16s) whose focus is primarily on science rather than on administration (they do not recruit, hire, review, or promote) and 4 division and department heads employed at the GS15 level.

Study participants were contacted in April/May 1996 and asked to participate in a self-report survey conducted by mail. In the original mailing they were also advised that they would be contacted at a later date to schedule a 90-minute follow-up interview. In May 1996, prospective interview participants were contacted by e-mail and voice mail and asked to participate in an interview.

*Editor's note: Although Asian Americans are not underrepresented in science and engineering, they were included in the interviews in order to gather information from a broad spectrum of ONR personnel. The phrase "minority males" as used in this Appendix, therefore, refers to all non-white males, including Asian Americans.

THE RESPONDENTS

Out of 53 members of the survey sample of program officers, 2 were out of the area and unavailable. Of the 51 remaining, 45 participated in interviews: 14 females, 9 Asian American males, 1 African American male, and 22 white males.

Twenty-six out of 27 members of the survey sample of senior executives participated in interviews. One senior executive declined to participate.

THE INTERVIEW PROCESS

The interviews of the program officers were conducted in confidence in a private office provided by ONR's Office of Human Resources in a different building from the one in which all but two of them worked. The interviews of the senior executives were conducted in the offices of the respondents, or, at their request, in nearby conference rooms. The interviews were tape recorded with the explicit understanding that a record was being made for reference purposes only, and that only the interviewer would listen to the tapes. Once the notes were complete, the tapes were destroyed. Three program officers and two senior executives asked that no tape recording be made. All interview tapes were taken off-site for playback.

The interview protocol (appended to this report) was constructed as an inductive or ethnographic interview based on 12 questions with multiple probes. The protocol included an informational interview calling for facts; an attitudinal interview inviting the expression of attitudes, opinions, and beliefs; and a critical incident interview calling for examples of behaviors or events illustrating best and worst experiences that (1) program officers attributed to their being women, minority males, or white males or that (2) senior executives related to race or gender. In addition, the senior executives were asked to respond to each question, not only by describing their own experience, but also by describing how they were using their leadership to promote diversity.

Respondents were articulate and cooperative. Although some interviews lasted only 30–45 minutes, many respondents took the time to discuss every issue at length, with the result that many of the interviews lasted 90 minutes. Several program officers returned a second time, and a number of program officers and senior executives stated that they were available to help if there was a further need for their input. As a result, more than 60 hours of tape-recorded interviews and notes with program officers and more than 40 hours of tape-recorded interviews and notes with senior executives were generated.

METHODOLOGICAL CONSIDERATIONS

- No generalization accurately describes every member of each group because of differences both among and within groups with regard to every issue addressed.

- There was greater heterogeneity in responses among the white male program officers than among female and minority program officers, making categorization and frequency counts of responses to the primary questions difficult.

- Senior executives often expressed multiple view points simultaneously (e.g., emphasizing the positives and the negatives associated with diversity), making categorization and frequency counts of responses to the primary questions difficult.

- The interviewing approach used was inductive. As such, it focused not on responses to questions that could be answered positively or negatively, but on eliciting the range of individual and organizational behaviors that influence the experience of diversity in the ONR work place.

- The interview was designed to capture the interviewee's perception of what needs to be changed or of what is discriminatory and what is not.

PART II: SUMMARY OF INTERVIEW RESULTS WITH PROGRAM OFFICERS

The context within which scientists and engineers work at ONR and the constraints imposed by this context should be kept in mind as findings are reviewed.

- ONR is the premier research funding organization of the Navy with a long and distinguished history of funding basic and applied research, especially at universities and federal laboratories.
- The S&E work force at ONR headquarters is comprised of senior scientists and engineers who serve both as science administrators overseeing multimillion-dollar funding programs and as scientific researchers conducting their own research programs.
- The many benefits associated with ONR include the opportunity to contribute to new knowledge by funding cutting edge research in a program of one's own design; the capacity to make funding decisions without peer review; high rank (more than two-thirds are employed at the GS15 or senior executive level); high salary (although many interview participants were quick to point out that salaries in industry are often higher); numerous opportunities for professional development; relative job security; and the elite status of belonging to the premier science and engineering research funding organization of the Navy.
- ONR reorganized approximately two years ago and merged with two other offices—the Office of Naval Technology (ONT) and the Office of Applied Technology (OAT). Since that reorganization, the focus has been on the integration of basic research with application to the fleet.

ATTITUDES TOWARD THE WORK ENVIRONMENT

- Women, minorities, and white males commented on the many opportunities for professional development and the flexible work-family arrangements.
- Women and minority males particularly praised the value ONR placed on professional ethics and scientific integrity.
- Members of all three groups described ONR as "the best job" that they have had, citing specifically their ability to fund promising new research and development and to conduct their own research.
- Women and minority scientists at ONR are sophisticated about diversity issues in the science and engineering work force. The Diversity Committee, initially appointed in December 1994, has analyzed the problems faced by female and minority scientists at ONR and proposed solutions.
- Women at ONR have formed chapters of WISE (Women in Science and Engineering) and WSEC (Women in Science and Engineering Council) to address issues of diversity. An invitation in 1994 to Nora Slatkin, Assistant Secretary of the Navy, to attend a luncheon organized by WISE resulted in action to improve the status of women and minorities at ONR.
- Many of the women referred to the differences between "the white male culture" and the "culture of minorities and women." The white male culture was described as hostile, aggressive, argumentative, competitive, and vocal. The culture of minorities and women was

described as cooperative, nurturing, quiet, and reflective.

- Some female Ph.D.s who have been at ONR for a number of years perceived themselves to be second-class citizens whose contributions are not valued. Rather than reporting the experience of being at ONR as positive or neutral, they described the environment as hostile. Some non-Ph.D. women with similar years of experience at ONR also reported dissatisfaction. Some women were interested in senior executive status but have had no opportunity to be promoted. Others were less interested in management than in making a difference as scientists or engineers.

- Women and minorities described ONR as "paying lip service to diversity," and several reported little evidence of serious intent to improve the status of minorities and women.

- White males described a very different ONR characterized by friendliness, camaraderie, and "bending over backwards to promote diversity," an organization in which the best candidates are recruited, hired, and promoted, regardless of gender or race and ethnicity.

- The experiences of white males appeared to be more variable than the experiences of women and minorities, making efforts to describe them in terms of broad generalizations more difficult. Not every white male was satisfied with his experience at ONR, although most described a far more sanguine climate than was reported by the women or minorities. For some—although by no means all—of the white males interviewed, promotion to senior executive status is a coveted prize for successful performance. Others, however, described little interest in managing people, as distinguished from overseeing research programs.

ATTITUDES TOWARD DIVERSITY

- Many males perceived that the pool of women and underrepresented minorities (African Americans and Hispanics, in particular) for positions at ONR was very small. Other male respondents reported that qualified women and minorities were out there, but that serious efforts to recruit them were not being made. There was a consensus that science and engineering education was a critical issue, especially for underrepresented minorities. In this connection, attention was called to ONR's HBCU/MI program.

- The women and men who are committed to diversity principles believed that the problems women in particular face will be solved only as

  - more women are actively recruited to ONR
  - more women are promoted or hired directly into senior executive positions
  - senior women (and men) take on formal roles as mentors to junior women
  - women take on visible senior roles representing ONR to the scientific and engineering community outside ONR
  - men are required to interact with female scientists and engineers in positions of authority

  Of these various activities, recruitment of women and mentoring appeared to be at the top of the agenda.

PART III: SUMMARY OF INTERVIEW RESULTS WITH SENIOR EXECUTIVES

ATTITUDES TOWARD DIVERSITY

There was a consensus among most senior executives that there is a need to increase diversity when it is defined as the recruitment of more women and underrepresented minorities to ONR. Although there was little awareness of the hostile, aggressive, or argumentative behavior reported by some female program officers, several senior executives acknowledged that they would not necessarily be aware of such behavior even if it did exist since staff were careful about how they behaved in their presence. Several did admit that the ONR environment was fast paced, competitive, and "bottom-line" driven, but they also stated that it was that way for men as well as for women. And several were emphatic that certain aspects of the ONR culture should remain the same, especially the use of an adversarial approach in recruitment interviews and in meetings.

Two sentiments predominated regarding the recruitment of diverse populations:

- The first saw diversity as strengthening ONR by bringing a variety of viewpoints to bear on the scientific and engineering enterprise itself. This group stated that diversity will advance the enterprise in a way that homogeneity of perception and experience cannot.

- The second feared that diversity would dilute the strength of ONR, particularly if it meant lowering standards of technical competence in order to achieve it.

Two broad approaches to achieving diversity in recruitment were discussed:

- The first reached out to encourage outstanding senior women and underrepresented minorities in academia, industry, and government to come to ONR on a permanent basis. Efforts in this direction have not succeeded to date.

- The second approach involved "growing one's own" by encouraging mid-level minority and female scientists and engineers from universities and the Navy laboratories to accept temporary positions as IPAs and detailees, with the idea of recruiting from this pool. It also involved encouraging new minority and female Ph.D.s to accept positions in Navy laboratories since it was perceived that they were less likely to want to take on such positions later on in their careers. This is the approach that ONR has been taking, although it has not produced the kinds of results that most senior executives expected. A variation of this approach involves bringing in senior female and minority scientists or engineers from academia and industry for a few months to occupy a special "chair" created for this purpose. This was recently tried for a first time, but with a senior white male university professor.

For the most part, the discussion focused on the recruitment of women rather than underrepresented minorities (e.g., African Americans and Hispanics). The consensus among most senior executives was that there is an exceedingly small pool of underrepresented minorities in the science and engineering fields of interest to ONR; that few members of these racial and ethnic populations enter technical fields; that the numbers in fields such as physics were, if anything, declining; and that ONR could not compete successfully with academia and industry or even other government agencies for the very small numbers of outstanding scientists and engineers drawn from these populations.

INTERVIEW GUIDE

Staff from the National Research Council will conduct interviews with a sample of ONR staff during December 1995 and January 1996. In preparation for these meetings, those to be interviewed have been asked:

(1) to consider what they perceive to be the formal policies and informal practices that may, inadvertently, perpetuate inequities in the experiences of scientists and engineers—male and female, minority and white, physically disabled and those with no physical disabilities; and

(2) to reflect upon their own experiences and to tell us whether they have ever experienced or witnessed events illustrating either especially equitable or fair policies and practices or differential treatment of scientists and engineers on the basis of gender, race/ethnicity, or physical ability.

We especially want to know how conditions, policies, and practices at ONR differentially affect those whose perspectives (and backgrounds) are different.

CONDITIONS, POLICIES AND PRACTICES

1. ***Understanding ONR as an Organization.*** Would you say that you have a good understanding of how ONR works (policies and procedures) formally and informally? How have you gotten this information?

2. ***Valuing Diversity in the Science and Engineering Work Force.*** What conditions, policies, and practices need to change in order to achieve diversity—i.e., increased representation of targeted groups (women, racial/ethnic minorities, and the physically disabled)—in ONR's science and engineering work force? What history and traditions may have perpetuated, perhaps unintentionally, a science and engineering work force that is predominantly white male, especially at senior management levels?

3. ***Access to ONR: Advertisements, Recruitment, and Hiring:*** How were you recruited for your position—for example, through an informal network, by a search committee, or in response to a position announcement?

   What were the terms and conditions of your original employment with ONR? Were these, as far as you know, competitive with those offered other ONR employees in your professional discipline (series)?

   What should be done to attract members of the three targeted groups to your field? What would <u>you</u> do to recruit them to ONR?

4. ***Annual/Periodic Review and Promotion:*** Do you know what you have to do to be promoted? Are there easily accessible mechanisms (e.g., ombudsperson, other dispute mediator, or written procedures) for formal or informal appeal of decisions related to annual/periodic review? Does ONR or its departments review the evaluation process periodically to determine that men and

women, minority and non-minority, are evaluated equally stringently (e.g., as measured by a study of similarities and differences in level of research, ratings or other indices)? Are you satisfied with the review and promotion process? If not, what changes would you recommend?

5. ***Professional Development:*** What opportunities exist at ONR for your professional development? Why have you (not) participated in ONR's Research Opportunities for Program Officers Program? What opportunities do you have for upward mobility? Are your opportunities less than, equal to, or greater than those of others? Why do you say that? Do restrictions in professional development opportunities contribute to a staffing pattern in which women and racial/ethnic minorities are concentrated in junior and mid-level management positions while men dominate senior management? If so, what do you believe can be done about this?

6. ***Work-Family Arrangements:*** How does ONR accommodate scientists and engineers who have family responsibilities? Are working conditions (e.g., long hours) forcing you to shortchange your family? What work-family arrangements would you like to see at ONR (e.g., flex time, part-time employment, child care on ONR premises, parental leave for both parents at childbirth or adoption, family leave, employee assistance with such needs as alcohol abuse and dependent care, including elder care)?

7. ***Discrimination:*** What are your experiences of especially positive treatment of women and minority scientists and engineers at ONR, attributable to equitable conditions, policies or practices of your workplace?

☐ No experiences

☐ Experience(s) described during interview
What happened?
Who was involved?
What led to this situation?
What was the outcome?
Why does this appear to you to be an example of especially equitable treatment?

8. A broad range of behaviors and events perpetuate inequities for women and minorities in the workplace. What are your experiences of sexual or racial discrimination at ONR? Are there formal and informal opportunities for dissent, mediation and grievance without reprisal? Would you describe any of your experiences as "critical incidents" affecting your motivation, achievement, or other attitudes or behavior? If you have one or more examples, please take a few minutes to describe each critical incident.

☐ No experiences

☐ Experience(s) described during interview
What happened?
Who was involved?
What led to this situation?
What was the outcome (positive or negative)?
Why does this appear to you to be an example of discrimination?
If you knew the situation was actionable, but your decision was not to take action, why?

9. Since joining the staff of ONR, would you say that overt discrimination based on gender, race/ethnicity, or other factors has:

- □ Decreased
- □ Increased
- □ Remained the same
- □ Discrimination does not occur
- □ Don't know

10. Since joining the staff of ONR, would you say that subtle discrimination based on gender, race/ethnicity, or other factors has:

- □ Decreased
- □ Increased
- □ Remained the same
- □ Discrimination does not occur
- □ Don't know

11. ***Ethics, Professional and Scientific Integrity:*** To what extent would you describe your working environment at ONR as ethnical, characterized by professional and scientific integrity? In responding, consider whether (1) research findings have ever been distorted, (2) scientific errors have been acknowledged or concealed, and (3) personal behavior has always been ethical and professional.

12. ***Physical Plant:*** Are you satisfied with the physical environment? Is the physical plant well maintained? Is physical space equitably allocated?

APPENDIX D

EXAMPLES OF SUCCESSFUL DIVERSITY INITIATIVES IN OTHER ORGANIZATIONS

AEROSPACE CORPORATION

Aerospace Corporation has made significant progress from 0.01 percent women in their technical workforce in 1962 to 12.5 percent women in 1993. Its recruitment of women is aided by the involvement of many women who act as mentors, by policies and practices that are attractive to women, and by many dedicated people who are working toward creating a positive climate that contributes to the successful recruitment of women.

Women's committees or networks within a company can be vitally important influences on the recruitment and retention of women, and the Women's Committee at Aerospace has been not only a leader for change, but also a strong support system that has helped women gain confidence. Issues addressed include the following:

- Developing a maternity leave policy
- Equalizing employee benefits for secretarial and technical staff
- Establishing awards for women
- Researching and making recommendations to the president and executive staff on female candidates for the board of directors

ALCOA

ALCOA Technical Center, the research facility of the Aluminum Company of America, is a leader in the industry in developing programs for attracting and retaining talented women. Effective policies and programs have been developed to help employees balance work and family responsibilities.

To establish recruitment programs and to develop university relations, ALCOA representatives were sent directly to select colleges and universities with large numbers of women and minority graduates in the sciences and engineering. The Pilot School Program is the result of such efforts, and it serves to establish meaningful technology linkages between the ALCOA coordinator and university faculty and students. Through its Minority Fellowship Program, ALCOA develops relationships with high potential female and minority candidates

APPLE

A Systemic Approach to Diversity Programs

Include those who will be involved or interested in the outcome of the program or intervention in developing programs and interventions that support diversity in all phases of the employment cycle—sourcing, recruiting, interviewing, orientation acculturation, coaching, and development/ promotion. Following is a list of what Apple is doing or is considering doing in each of these areas.

Sourcing. Establish ongoing relationships with professional associations and universities. For example, Apple engineers act as "professor for a day" and/or set up and support computer labs on campus, or they might host product demonstration booths at professional conferences and meetings and arrange for senior scientists and managers to keynote functions.

Recruiting. Campus recruiting, professional networking, and word-of-mouth. Provide bonuses to individuals who recommend new hires for hard-to-fill positions and offer internships.

Interviewing. Provide training to support cultural diversity in the process (e.g., how to accurately assess a candidate's qualifications across cultures, and how to appropriately and comfortably communicate one's qualifications). Inform candidates about what to expect during the interview process and monitor the interview and selection processes.

Orientation/Acculturation. Start up a buddy program and an employee association.

Coaching. Mentoring.

Development/Promotion. Rotate assignments; Leadership in Action, a junior/senior partnership program; "Bridges—Managing a Diverse Workforce" training program; and roundtable discussions about current business and career issues featuring company leaders.

Climate. Cultural celebrations and diversity awareness training ("A Winning Balance").

AT&T

AT&T Bell Laboratories "value both women and men employees who are well grounded in their technical expertise and who have an understanding of the business world." AT&T is committed to encouraging this broader base of knowledge and to improving the working climate for its female employees.

AT&T has developed, collaboratively with human resources and technical staff, a University Relations Summer Program and offers scholarships and fellowships such as the Engineering Scholarship Program, the Cooperative Research Fellowship Program, and the Graduate Research Program for Women. Another effort at AT&T is sponsorship (with the National Science Foundation and other U.S. companies) of the Women in Engineering Program Advocates Network (WEPAN). WEPAN includes efforts to increase the numbers of women studying engineering, to establish mentoring and summer internships involving the private sector, and to find positions for women engineering graduates. WEPAN also works with women in professional positions.

The Employee Counseling Service is a highly effective program that provides private and confidential one-on-one counseling sessions, consults with management about special organizational and individual issues, and presents special seminars and workshops on subjects such as balancing work and family, single working parents, and dual-career couples.

CORNING, INC.

The Corning Professional Women's Forum: Networking—A Guide for Getting Started

A voluntary network was created to foster professional growth for women employees of Corning. Five basic kinds of networks were identified:

- Professional/trade group
- Mixed industrial women's business network
- Mixed network of men and women
- Volunteer network
- Strong network of family and friends

A three-fold approach was recommended to structuring the network:

- **Buddy system.** Know someone who has your best interests at heart.
- **Board of Advisors.** Know a group of people who can provide advice.
- **Resource Network.** Work to expand the network to include a variety of people and levels.

United States Department of Energy

The United States Department of Energy (DOE) views people as its most important resource and values the needs of individuals. It is committed to attracting and accommodating women as part of the diverse work force and places the safety, health, and well-being of the work force as the highest priority. DOE is helping women achieve their professional goals through targeted recruitment and employment programs.

- **Upward Mobility** (a government-wide program) is a competitive recruitment option enabling DOE management to select highly qualified, lower-level employees for trainee positions.

- **Management Intern Development Program and Career Entry and Development Program** offer recruitment opportunities to entry-level professionals for technical, administrative, and managerial jobs.

- **Presidential Management Intern Program** matches the interest of outstanding graduates desiring public service careers in policy analysis and program management with the needs of federal departments and agencies.

- **Technical Leadership Development Program** (within the Office of Defense) and **Technical Manager Development Program** (within the Office of Environmental Management) are designed to select qualified graduates and to develop their technical management skills.

Many issues fundamental to the advancement of women in the work place revolve around the role of women as caregivers for children and elders and as providers of family economic stability.

- The **Eldercare Program** offers assistance and resources to employees taking care of older relatives.

- The government-wide **Federal Women's Program** was initiated 1980. The program at headquarters has included educational programs, reports, outreach, and activities exhibits at professional meetings. It also offers the opportunity for participation in the Federal Women's Program Advisory Council. In addition, each DOE field organization has a Federal Women's Program Coordinator and conducts numerous activities which include various role model/mentoring programs.

Education and training programs offered by DOE include federal and laboratory initiatives that advance the status of women by offering them opportunities to enhance leadership skills and to gain exposure to science and mathematics-based professions.

- Federal programs provide management training and greater access to management positions and include the **Professional Skills Training Program**, the **Women's Executive Leadership Program**, the **Leadership Enhancement Program**, and the **Senior Executive Service Candidate Development Program**.

- Laboratory initiatives are designed to stimulate interest in science careers through hands-on experience and include the **Pre-Freshman Enrichment Program**, the **Committee on Women in Science and Engineering**, and the **Summer Student Training Program**.

XEROX

Xerox Corporate Research and Technology (CRT) is working towards making Xerox Corp "the employer of choice for women in minorities by the year 2000." To meet its goal, a "Women's Council" was formed in 1991 to provide advice to senior management on issues related to recruiting and retaining women.

Council members identified five problem areas:

- the number of women in corporate research
- career development
- salary equity
- working environment
- benefits

As a result, the following recommendations have been made:

- Improve hiring and promotions
- Establish linkages with universities
- Create lateral hiring opportunities into other units of Xerox
- Allow cross-laboratory promotions

Through the formation of the "Women Managers Roundtable," the company is introducing a change in culture from the top.

APPENDIX E

COMMITTEE TO STUDY DIVERSITY IN THE SCIENTIFIC AND ENGINEERING WORK FORCE OF THE OFFICE OF NAVAL RESEARCH

BIOGRAPHICAL SKETCHES

Harrison Shull, *Chair.* Member, National Academy of Sciences; Member, Naval Studies Board. Dr. Shull has served in a variety of academic positions, including Chancellor, University of Colorado at Boulder; Vice President of Academic Affairs and Provost, Rensselaer Polytech Institute; and Provost and Academic Dean, Naval Postgraduate School, from which he has recently retired. He maintains membership in the American Association for the Advancement of Science, American Physical Society, American Academy of Arts & Sciences, American Chemical Society, as well as foreign membership in the Royal Swedish Academy of Science and the Royal Uppsala Academy of Arts and Sciences. Dr. Shull holds a Ph.D. from the University of California, Berkeley.

Linda Cain, Director, Office of University and Science Education, Oak Ridge National Laboratory. Dr. Cain has extensive experience with science teaching at the pre-college level and with the management and oversight of scientific careers through and beyond the college and graduate school levels. Her professional affiliations include the American Association for the Advancement of Science, American Chemical Society, Association of Women in Science, U.S. Department of Energy Precollege Evaluation and Assessment Committee, and the National Science Teachers Association. Dr. Cain has an Ed.D. from the University of Tennessee, Knoxville.

George Campbell, Jr., President and CEO, NACME, Inc. A Fellow of the American Association for the Advancement of Science, his many awards include the George Arents Pioneer Medal in Physics, the Drexel University Centennial Medal, and, on behalf of NACME, the 1996 Presidential Award in Science, Mathematics and Engineering Mentoring. Formerly an AT&T Bell Laboratories scientist and manager, Dr. Campbell has been a U.S. delegate to the International Telecommunications Union and spent two years on the Nkumbi International College faculty in Zambia. He has served on the U.S. Secretary of Energy's advisory board and is currently a member of the Board of Trustees and Executive Committee, Rensselaer Polytechnic Institute; the Engineering Advisory Council of Cooper Union; the Board of Trustees and Executive Committee, New York Hall of Science; the Advisory Board of the Merck Institute for Science Education; and the Board of Directors, Oak Ridge Associated Universities. He holds a Ph.D. in physics from Syracuse University.

Katharine B. Gebbie, Director, Physics Laboratory, National Institute of Standards and Technology. She is a Fellow of the American Physical Society and recipient of a Women in Science and Engineering (WISE) Lifetime Achievement Award. Dr. Gebbie serves on the Executive Committee of the American Physical Society's Division of Atomic, Molecular and Optical Physics; the Society's Committee on the Status of Women in Physics; and the Department of Energy's Fusion Energy Advisory Committee. She also represents the United States on the International Committee for Weights and Measures. Dr. Gebbie has a Ph.D. from University College London.

Eve L. Menger, Director, Characterization Science & Services, Corning. Inc. and Presidential Appointee to the National Science Board. For the last four of her 10 years at AlliedSignal, she was Senior Technical Officer. Dr. Menger has held academic positions at the University of Virginia (Vice Provost for University-Industry Relations and Professor of Chemistry) and the University of California, Santa Cruz. Her professional and public service affiliations include membership on the boards of directors of Science Service Inc., Hardinge Inc., Regional Science & Discovery Center, Literacy Volunteers, and Ceramics Corridor; and membership on the advisory committees of the University of Texas at Dallas, Doctor of Chemistry Program; Cornell Biotechnology Center for Advanced Technology; Oakridge National Laboratory,

Metals & Ceramics Division; and the Center for Photoinduced Charge Transfer at the University of Rochester. She is a member of the American Chemical Society, Committee for Chemistry and Public Affairs. Dr. Menger received her undergraduate degree from Carleton College and has a Ph.D. from Harvard University.

Herbert Rabin, Director, Engineering Research Center; Professor; Associate Dean, A. J. Clark School of Engineering, University of Maryland College Park; Member, Naval Studies Board. He has had prior experience as a senior scientist and manager at the Naval Research Laboratory and has served as Deputy Assistant Secretary in the Office of the Secretary of the Navy. Professional memberships include the American Association for the Advancement of Science, American Physical Society, Optical Society of America, and the American Institute of Aeronautics and Astronautics. Dr. Rabin has a Ph.D. in physics from the University of Maryland and serves as a Trustee of the National Technological University.

Frank J. Talamantes, Professor of Biology, University of California, Santa Cruz. He is a recipient of the British Transatlantic Medal Lecturer and of The Society for the Study of Reproduction Research Award. As a Fellow of the American Association for the Advancement of Science, Dr. Talamantes holds membership in the Endocrine Society, Society for the Study of Reproduction, Society for Biochemical and Molecular Biology, and The American Physiological Society. His research is in the area of biochemical endocrinology, which is funded by NSF, NCI, and the NIH. He is also actively involved in various committees and programs associated with the advancement of ethnic minority students in the sciences. He received his Ph.D. in endocrinology from the University of California, Berkeley.

Toni Tomacci, Director, Human Resources, MatriDigm Corporation; Adjunct Faculty Member, University of San Francisco and Dominican College. For the last 14 years, Dr. Tomacci worked for Apple Computer, Inc. where, from 1989 to 1996, she held the position of Diversity Consultant. She serves on the Board of Trustees, Pacific Graduate School of Psychology; the Corporate Advisory Board, Stanford University Institute for Research on Women and Gender; and the Vocational Education Advisory Council, City College of San Francisco. Her professional memberships include the American Association of University Women, the Association of Governing Boards of Universities and Colleges, Bay Area Organization Development Network, International Network of Women in Technology, Northern California Diversity Business Forum, and World Business Academy. Dr. Tomacci holds a doctorate in organization and leadership from the University of San Francisco.

Stephen J. Lukasik, *liaison*, Committee on Women in Science and Engineering. He has been Director of the Defense Advanced Research Projects Agency and Chief Scientist of the Federal Communication Commission. Dr. Lukasik has also held executive positions as Xerox, RAND Corporation, Northrop Corporation, and the TRW Space and Defense Sector. He holds membership in the American Association for the Advancement of Science, American Physical Society, Sigma Xi, and Tau Beta Pi. Dr. Lukasik received his Ph.D. in physics from the Massachusetts Institute of Technology and is a Trustee of Harvey Mudd College.